Going Ape

UNIVERSITY PRESS OF FLORIDA

Florida A&M University, Tallahassee
Florida Atlantic University, Boca Raton
Florida Gulf Coast University, Ft. Myers
Florida International University, Miami
Florida State University, Tallahassee
New College of Florida, Sarasota
University of Central Florida, Orlando
University of Florida, Gainesville
University of North Florida, Jacksonville
University of South Florida, Tampa
University of West Florida, Pensacola

University Press of Florida
Gainesville · Tallahassee · Tampa · Boca Raton
Pensacola · Orlando · Miami · Jacksonville · Ft. Myers · Sarasota

Going Ape

Florida's Battles over
Evolution in the Classroom

BRANDON HAUGHT

A Florida Quincentennial Book

Copyright 2014 by Brandon Haught

An early version of chapter 9 originally appeared in *Reports of the National Center for Science Education* 28, no. 4 (2008).

Printed in the United States of America on recycled, acid-free paper

This book may be available in an electronic edition.

19 18 17 16 15 14   6 5 4 3 2 1

Library of Congress Cataloging-in-Publication Data
Haught, Brandon.
Going ape : Florida's battles over evolution in the classroom / Brandon Haught.
pages cm
Summary: In this book, Haught chronicles the war over teaching evolution in Florida's schools, from the first shouts of religious persecution and child endangerment in Tallahassee in 1923 to the forced delays and extra public hearings in state-level textbook adoptions today.
Includes bibliographical references and index.
ISBN 978-0-8130-4943-4 (alk. paper)
1. Evolution (Biology)—Study and teaching—Florida. 2. Evolution (Biology)—Philosophy. 3. Creationism. 4. Religion and state. 5. Religion and science. I. Title.
QH362.H38 2014
576.8—dc23
2013044108

University Press of Florida
15 Northwest 15th Street
Gainesville, FL 32611-2079
http://www.upf.com

To my son, Caleb, my anchor in a stormy sea

Contents

Preface

I believe that the doctrine that man
is descended from a beast, a doctrine
unsupported by any scientific fact, and
directly contrary to the Bible account of
the creation, is the greatest menace facing the church today.

William Jennings Bryan, 1923

"As a trained biologist I never even realized that there was anything controversial about evolution," said Dr. C. Francis Byers, recalling his time as a new biology professor in the 1920s at the University of Florida in Gainesville. "It came as a surprise to me to be suddenly, as a young instructor, introduced into an environment where it could be even a fighting word, let alone a dangerous one."[1]

Byers had witnessed the genesis of a war over the teaching of evolution in Florida's schools that has lasted ninety years and still engulfs the state with undiminished strength. Over the years it has featured some of the most tenacious culture warriors imaginable. Many governors, state legislators, school board members, and teachers have endured raucous battles in the local and national spotlights, some by choice and others unwittingly.

But the people who sparked the conflicts were frequently average, passionate citizens driven by a sense of moral duty. "Evolution implies there is no need for a creator," said a grandmother and textbook watchdog in the 1980s, "and the high rates of suicide show our children's hopelessness." Her efforts forced delays and extra public hearings in state-level textbook adoptions. The teaching of evolution "is grievously offensive and inexcusably reprehensible to Bible believers," said a retired minister whose unrelenting activism started in the 1970s and lasted twenty years. At the time, he was responsible for prodding two school boards into mandating the teaching of scientific creationism alongside evolution. These sincere themes of religious persecution and child endangerment still echo forcefully across the state and will undoubtedly ring on for years to come. To these anti-evolutionists, nothing less than the future of the country is at stake.

A 2008 anti-evolution effort culminated in a standing-room-only chamber as the Florida Board of Education debated the merits of a new set of science standards that would determine what subject matter would be taught in science classrooms across the state. Despite the broad range of science subjects found in the tome before the board members, Sir Isaac Newton, the periodic table of the elements, and lab safety never came up. Instead, a single subject from a single branch of science got all of the attention: evolution. Several prominent figures took turns standing before the rows of television cameras and polarized, vocal spectators, arguing for or against evolution's prominence in the standards. But the standard-bearer for the anti-evolution side didn't have the credentials of a politician or a scientist or a movie star. She was a mom who felt strongly enough about her children's education to take on the Goliaths she saw around her. "All I want to do is have my voice heard," she said.

Reporters and activists frequently compared the 2008 showdown to the so-called trial of the century—the 1925 Scopes "Monkey Trial" in Dayton, Tennessee. However, no one referenced the gubernatorial debates, textbook protests, or local school board conflicts that had regularly popped up over the course of the twentieth century, year after year, in the Sunshine State. Those events flashed brilliantly in newspaper headlines but vanished just as quickly, rarely to be mentioned again. This book aims to bring those events back to life. It will show where Florida's war with evolution education started, identify its principal warriors, explain how the opposing sides' tactics have changed over time, and offer a glimpse of how the conflict is likely to unfold in the future.

Why focus on Florida? Why not Kansas or Tennessee or Louisiana? Simply put, Florida is a reflection of the whole country. There's good reason why Florida is considered a swing state in presidential elections. It's a lively mix of young and old, liberal and conservative, rich and poor, a smorgasbord of national opinions condensed into one state.

Many excellent books have documented in great detail the epic clashes that have happened across the country since the 1920s over the teaching of evolution. But the countless skirmishes that never made it to the U.S. Supreme Court or other prominent battlefields have been, sadly, overshadowed. As a result, small but crucial nuances in the overall debate are at risk of being forgotten. By shrinking the spotlight to one state, we can witness the ebb and flow of the national evolution war in microcosm. The raucous yet ultimately unsuccessful pushes to get creationism taught at local school board levels are just as fascinating and informative as the successful ones. Political candidates using the creationism/evolution conflict to club each other during campaigns create compelling dramas. A lot can be learned from watching a concerned parent grow into the role of a passionate, vocal activist.

They might not have realized it, but Florida's many anti-evolutionist leaders throughout the decades could trace their activist heritage to one of the war's most famous crusaders: transplanted resident William Jennings Bryan. His name is eternally entwined with the Scopes trial where he was a star attraction on the team that prosecuted teacher John Scopes in 1925 for violating a state law forbidding the teaching of evolution.

But before Bryan boarded the train that carried him north for what would be his final crusade, he was the most prominent anti-evolution agitator in Florida. It wasn't just a publicity stunt; it was a cause he truly believed in. "I was convinced, after talking to Bryan himself about this, that Bryan was absolutely intellectually honest," Dr. Byers recalled. "This wasn't a gimmick with him. And when he showed posters of a good-looking girl in the upper left-hand corner of the poster, let us say, and a chimpanzee or primate down in the lower right-hand corner and banner across it saying, 'Did your daughter come from this?' he really meant it."[2]

But as you are about to learn in this book, William Jennings Bryan was merely the beginning.

Going Ape

"Between the Devil and the Deep Blue Sea"

After three unsuccessful bids to gain the U.S. presidency, William Jennings Bryan and his wife, Mary, bought land in Coconut Grove in 1912 and had their first Florida home, Villa Serena, built there. It was intended as a winter retreat for Mary's health, but when Bryan resigned his post as U.S. secretary of state in 1915, the couple settled in year-round. Six years later, he made Florida his legal residence.

The former presidential candidate and famous orator quite naturally was a popular public local figure. He entertained U.S. presidents, foreign dignitaries, and other notables of the time at his home. This Florida "retirement" was more like setting up a new headquarters for his still-energetic life. He made a living giving public lectures across the state on a wide variety of subjects, and his Tourist Bible Class in Royal Palm Park in Miami attracted thousands of people Sunday mornings December through April for a decade.[1] He was in high demand by organizations hoping to take advantage of his popularity and gifted public-speaking skills to promote their own causes.

Bryan's crusade against the teaching of evolution as a fact started in 1921 after he had heard from many concerned parents who attended his religious speeches that their children in college were turning away from biblical teachings. Bryan already had a wary eye on evolution's claims

William Jennings Bryan conducting a Bible class in Royal Palm Park, Miami, 1921. (William A. Fishbaugh, State Archives of Florida.)

about human origins, which he thought obviously conflicted with the Bible's creation account, but for years he hadn't felt the need to engage the enemy. However, now the dangers of teaching evolution as if it were an established fact seized Bryan's attention.

He attributed the spark of World War I and other conflicts around the world to the heartless competition of Darwinism. Children and young adults were being led morally astray by the materialistic philosophy that evolution implied. He understood evolution to be based on the principle of the strong dominating and wiping out the weak, which he referred to as a "law of hate." This stabbed right through the heart of Bryan's devout Presbyterian beliefs.

Another factor driving his anti-evolution efforts, and a big reason why he took on the Tennessee case, was his belief that public school teachers were employees of the communities in which they worked. If the parents in those communities didn't want subversive subjects like evolution taught, then teachers had better comply. He argued that

Darwinists were clearly in the minority, and their beliefs shouldn't override the rights of the majority.

At about the same time Bryan started focusing on this perceived danger, anti-evolution sentiments were bubbling up elsewhere in Florida. For instance, in May 1922 the Southern Baptist Convention met in Jacksonville, and among its members' priorities was the desire to see science textbooks "free from erroneous statements on evolution." The Baptists made it clear that "no man can rightly understand evolution's claim as set forth in the textbooks of today, and at the same time understand the Bible."[2]

That same year, the Miami Christian Council, organized by members of several Miami churches, heard a report during a regularly scheduled meeting that "the county school board had conferred with the teachers" and that "no evolution or Darwinian theory would be taught in Dade county public schools." A *Miami Herald* editorial had a sharply sarcastic reaction: "Our young people ought by no means to be allowed to see the *Miami Herald*, at least until the school board has carefully gone over it and cut out its dangerous features, for the *Herald* occasionally makes mention of ships that are sailing around the world, which is manifestly impossible, and this paper even mentions favorably, once in a while, the theory of evolution."[3]

However, the *Herald*'s views didn't reflect those of the majority of the citizens. Bryan easily tapped into the mounting anti-evolution fervor, committing to it with all of his considerable energy. Nothing was going to distract him. He turned down a committee chairmanship in the Presbyterian Assembly because he preferred to stay in an education committee where he could devote all his time to the fight "against the Darwinian theory." Bryan said, "I believe that the doctrine that man is descended from a beast, a doctrine unsupported by any scientific fact, and directly contrary to the Bible account of the creation, is the greatest menace facing the church today."[4]

"It Is Better to Trust in the Rock of Ages"

Bryan easily made friends in the Florida legislature. As he declared war on evolution, he prodded the lawmakers to take up arms with him, which they gladly did. This resulted in House Concurrent Resolution No. 7, presented April 18, 1923, by Rep. S. L. Giles of Franklin County:

> WHEREAS, the Constitution of the State of Florida expressly states in Section 6 of the Declaration of Rights that, "No preference shall be given by law to any church, sect or mode of worship, and no money shall ever be taken from the public treasury directly or indirectly in aid of any church, sect or religious denomination, or in aid of any sectarian institution."
> And,
> WHEREAS, the public schools and colleges of this State, supported in whole or in part by public funds, should be kept free from any teachings designed to set up and promulgate sectarian views, and should also be equally free from teachings designed to attack the religious beliefs of the public, therefore, be it
> RESOLVED by the House of Representatives, the Senate concurring, That it is the sense of the Legislature of the State of Florida that it is improper and subversive to the best interests of the people of this State for any professor, teacher or instructor in the public schools and colleges of this State, supported in whole or in part by public taxation, to teach or permit to be taught atheism, agnosticism, Darwinism, or any other hypothesis that links man in blood relation to any other form of life.[5]

Bryan had wanted an actual law rather than this non-binding resolution. Enthusiastic yet skittish legislators disagreed, wishing to stay out of the line of fire of any potential lawsuits a law could inspire.[6] Nonetheless, Bryan was satisfied.

When the resolution was introduced, the House members agreed "that the rules be waived" so that it could be officially adopted right away without having to wait for the usual further consideration and debate on a later date. The official *House Journal* didn't note any opposition to this fast-tracking move.

Bryan made a personal appearance before the legislature on May 11 to lend his support and suggest a few changes to the resolution. He gave a nearly two-hour speech to a joint session of representatives and senators. Acknowledging that Christianity couldn't be promoted in the classrooms due to constitutional prohibitions, he asked that "you do not allow the atheists, agnostics, or the Darwinists to spread their doctrine there."[7]

Bryan suggested that the words "or to teach as true" be slipped in before the word "Darwinism" in the resolution. The "or to teach as true" phrase reveals something interesting about Bryan's feelings on the evolution subject. He accepted the "day-age" creation theory, which meant he believed that when the Bible outlines what was created on each day, those days could have actually lasted millions of years each.[8] Evolution might not be all bunk, either, Bryan thought. It might apply to other living things, just not humans. He had no problem with the subject being taught as "a theory," meaning he considered evolution to be just a guess or unsupported idea. But he did have a big problem with its being taught as a factual explanation for human origins.

Bryan was realistic and calculating when advocating for the resolution. He felt that establishing a punishment for teaching evolution would rouse too much opposition and threaten its passage, so he recommended against it. Another reason for not wanting repercussions was that he felt the resolution was directed at "an educated class that is supposed to respect the law."[9] Besides, if the law is not obeyed, penalties could be imposed during the next legislative session, he thought.

The day after Bryan's presentation, the House decided to revisit the already-approved resolution. They quickly amended it with Bryan's suggestions and re-adopted it.[10] The Senate then fast-tracked its approval just as the House had done. Thus, without any fuss or drama, the Florida legislature had officially recommended to the state's educational institutions that evolution not be taught.

To Bryan's consternation, though, the resolution was largely ignored. Newspapers didn't give it much attention, and so most people didn't even know about it. Bryan was forced to remedy that on his own as best he could. The *Tampa Morning Tribune* ran a story in June 1923 that told of Bryan's frustration with a *Chicago Tribune* editorial that took him to task for his particular interpretation of the Bible. Bryan fired back a response that was published in the newspapers, using the resolution as ammunition. "My views are set forth in a resolution unanimously adopted by the legislature of Florida," he wrote. "Please note that the objection is not to the teaching of evolutionary hypothesis as an hypothesis but to teaching of it as true or as a proven fact."[11]

Bryan's network of like-minded allies in Florida wasn't restricted to lawmakers. He was introduced to University of Florida, Gainesville, president Albert A. Murphree in 1916, and they quickly discovered that they had similar religious views. Murphree's moral convictions were reflected in his desire for students to attend chapel services a few times a week and his refusal to employ any professor "who is known to be irreligious." In 1919 he asked Bryan for help raising funds for a student religious activities building, and Bryan threw himself into the cause. A couple of years later, Bryan enlisted Murphree's aid in launching a nationwide campus campaign against alcohol. Faculty and students were asked to sign a "whiskey pledge" that stated, "We, the undersigned, promise, God helping us, never to use intoxicating liquor as a beverage."[12]

In 1921 Bryan published the pamphlet *The Menace of Darwinism*, in which he said "Darwinism is not science at all; it is guesses strung together" and "It is better to trust in the Rock of Ages, than to know the age of the rocks; it is better for one to know that he is close to the Heavenly Father, than to know how far the stars in the heavens are apart."[13] He gave Murphree two hundred copies to distribute on campus.

Bryan also gave a speech in February 1922 to the university's student body, hoping to counteract the effects that learning about evolution was having on them. Evolutionists had failed to produce evidence of one species changing into another, he said. He warned students of the "real evil" of Darwinism, namely, calling into question the literal interpretation of the Bible. Bryan was under the impression that Murphree was on his side in this matter, praising him for being a leader "of a great university who has not been carried away by the ape theory."[14] He returned to the university in 1924 to give several more talks on a variety of subjects, including evolution. It turned out that while Murphree appreciated Bryan's oratory skills and called him a "great master" of public speaking, he disagreed with Bryan's anti-evolution stance. After the 1924 evolution speech, Murphree noted: "His foolish notions on evolution came in for a round, though he was not so bitter and not quite so unreasonable."[15]

Dr. C. Francis Byers, a biology professor at the University of Florida in Gainesville during the 1920s, had firsthand confirmation of Murphree's pragmatic views on evolution. He recalled a conversation they had about how to weather the storm when the state legislature had approved the 1923 anti-evolution resolution. Murphree explained to Byers that he would comply with whatever the law directed, since the school was state-supported. On the other hand, Murphree knew that the professors considered evolution a vital part of their instruction.

As Byers and Murphree talked, they worked out a compromise. Instead of using the word "evolution," professors would refer to the concept as "progressive development." Murphree was OK with that, but he wanted to know if human "progressive development" would be taught. Byers said yes. According to Byers, Murphree replied, "Well, I wish that you wouldn't stress that part." Byers understood, and a deal was struck. "So that's the way it was done," Byers said. "There was not real trouble here. Nobody that I know of got arrested and nobody was kicked out of the university that I know of for involving himself in the evolution thing."[16]

Dr. John Henry Davis, another biology professor at the University of Florida, also recalled that the evolution storm didn't really blow through the classroom that much. "That was an issue that was brought up, but never taken below the level of the present heads of the department," he said. "We had decided that the less we talked about it, the better. We just went on and taught it."[17]

For Bryan and other ardent anti-evolutionists, that was a problem. They saw that the state legislature's resolution wasn't being taken seriously, so they decided to do exactly as Bryan had suggested earlier: pass new legislation with some bite to it. Their next opportunity came during the 1925 legislative session. However, Bryan's influence was notably absent, since he had departed for the Scopes trial and then died shortly afterward while still in Tennessee.

Despite the loss of Bryan, Florida's anti-evolutionists were still going strong. Reps. Albert W. McLeran of Suwannee County and Shelton Philips of Levy County introduced the following bill in 1925:

House Bill No. 691:
A bill to be entitled An Act prohibiting the teaching of the Evolution Theory in all Universities, Normals, and all other public schools of Florida, which are supported in whole or in part by the public school funds of the State, and to provide penalties for the violation thereof.[18]

The bill was referred to the House Committee on Education. Members there tweaked the language so that the bill then read:

> A bill to be entitled An Act to prohibit the teaching in any college, university, normal school or other school in this State, supported in whole or in part by the public funds of the State or any subdivision thereof, as fact, any theory which denies the divine creation of man, and to provide penalties for the violation thereof.[19]

The bill wasn't brought up again during the session, and it died. However, spurred on by the nationwide publicity the Tennessee trial had attracted, the teaching of evolution in Florida's schools was under more scrutiny that ever.

"A Storm of Indignation Has Been Aroused"

The great battle of the age was now on between Christianity and evolution, or so said George Washburn.[20] He was one of the many men who heard the call to pick up the anti-evolution fight where Bryan had left off upon his death. Drawing upon the wealth he had accumulated from dealing in real estate and hotels, Washburn helped establish the Bible Crusaders of America in 1925. God had chosen him for this task, he said, and he took on the job with gusto from his Clearwater headquarters.

The Bible Crusaders' slogan was simple and unambiguous: "Back to Christ, to the Bible, and to the Constitution."[21] The Crusaders felt that there was a growing "modernism" trend sweeping across the country and that Darwinism was at its root. Part of Washburn's vision was to sweep this un-Christian evolution out of the nation's schools. One of the first steps, though, was to start locally; evolution needed to be removed from Florida's textbooks.

Washburn was probably beside himself with what he saw in Florida's main biology textbook at that time: Benjamin C. Gruenberg's *Elementary Biology*, published in 1919. Evolution was featured unapologetically throughout. Page 483 offered a full-page evolutionary tree illustration with a man in a suit pictured at one branch with a bat and an elephant on related branches nearby. A chapter devoted to the various lines of evidence for evolution ended with the summary: "There are many attempts to explain how evolution is brought about, and these may be called theories of evolution; but as to the fact of evolution biologists are in substantial agreement."[22] Gruenberg was even one of the witnesses the Scopes defense had wanted to testify on behalf of evolution's validity at the Tennessee trial. Washburn vowed that this book had to go.

The spark of Washburn's campaign can be attributed to the massive changes going on in society in the late nineteenth and early twentieth centuries. Following World War I, a clash developed between rural-agrarian and urban-industrial ways of life. There was a steady and irresistible drive by politicians and businessmen to industrialize and advance, but rural communities saw this as an attempt to rob them of control over their own lives.[23]

Education reform, spurred by progressive ideologies originating in the cities of the North, was a big part of this change. Schools that had been controlled entirely by communities at the local level were now becoming centralized at the district and state levels. School attendance was now compulsory, because it was "viewed as the key to cultural progress."[24] This met determined resistance from rural families, who felt that unchecked modernization was ripping the heart out of long-standing tradition and family values.

Companies headquartered in the big cities published the textbooks used nationwide. The authors typically had a social reform agenda, and their writing focused on civic responsibilities. The biology texts of the time steered lessons toward showing how biology applies to modern concerns like sanitation, a healthy diet, and the effects of alcohol. For instance, in Gruenberg's *Elementary Biology* the chapter "Digestive System in Man" was followed by chapters on "The Social Side of the Food Problem" and "Health and Food Standards." The chapter on "Wastes and By-products of Organisms" led to chapters on "Hygiene of Excretion," "Excretion and Fatigue," and "Fatigue and the Worker." These subjects were often heavily skewed toward city living.

Some textbook editions were targeted specifically at rural schools, but purchasing decisions were made by committees at the state level, and those groups were dominated by representatives from the larger cities. They weighed the options and typically went with the urban-leaning textbooks. The publishing companies also had their bottom lines to consider, motivating them to produce fewer editions for a higher profit. As historian Adam Shapiro notes, "Textbook salesmen could, and did, lobby a state or school board to change its curriculum to suit available books."[25] As a consequence, if something was in the textbooks it was taught in the classrooms. Enter evolution.

Evolution actually wasn't something new to textbooks; it was a matter-of-fact concept that had been in textbooks for several years. But now textbooks that unapologetically discussed evolution were landing with a loud thump in school districts where local leaders were already feeling steamed about what they saw as the state government meddling in local affairs. That anger was about to find a symbolic outlet. Despite Washburn's and the Bible Crusaders' fiery rhetoric, the conflict wasn't just about science versus religion. It was about control, the vision of the future, and personal values.

The Bible Crusaders poured relentless pressure on state politicians, saying that "a storm of indignation has been aroused among a majority of the voters" because evolution was in the state textbooks. Washburn relied heavily on one of Bryan's main arguments: separation of church and state. Since Christian Bible study wasn't permitted in the public schools, the "false religion" of evolution shouldn't be allowed either. He demanded that Governor John Martin and his cabinet, who served as the state's board of education, take immediate action. "We will organize and marshall the tremendous voting strength in every state, and with such potential energy and dynamic force behind us, we will make and unmake governors and other state officers, as well as congressmen and other statesmen, on this text-book question," the Crusaders announced. They demanded that evolution either be taken out of the current books or that the books be discarded entirely. "In this demand, we shall be supported by 80 per cent of the voters of this state, because, with them, the fundamental principle, that no religion, false or real, should be taught in the public schools, is stronger than friendship or party allegiance."[26]

The Crusaders denied that they were against all science instruction; rather, they were "against teaching of unproved guesses."[27] The anti-evolution resolution passed by the state legislature two years prior was Washburn's foundation. He explained to W. S. Cawthon, the state superintendent of public instruction, that it had stemmed from the constitution's prohibition of teaching religion in schools supported by taxes.

Fortunately for Cawthon and the governor, there was an easy way out of this conflict. Gruenberg had authored a new textbook that came out just as the Scopes trial was about to kick off. *Biology and Human Life*, avoided any mention of evolution. Textbook publishers had been spooked by the Scopes trial. Spotlighted during that trial was George William Hunter's textbook *Civic Biology*, and the negative publicity generated really upset its publishers. A few years earlier, Gruenberg's publisher saw the coming conflict and asked Bryan directly what would make a biology textbook acceptable and inoffensive. Bryan's answer was simple: "eliminating the objectionable phraseology [evolution] from the text books." The result was *Biology and Human Life*, which replaced *Elementary Biology* as Florida's textbook of choice. Washburn was ecstatic. In a letter to Tennessee governor Austin Peay on March 11, 1926, he gloated over what he called, "Victory number one."[28]

Now that the Bible Crusaders had notched a win, anti-evolutionists grew bolder. There were rumors that 1927 would see yet another attempt in the state legislature at outlawing evolution in schools. This prompted a preemptive opposition strike before the bill was even filed, according to the *Tampa Tribune*. "The desire that the Legislature of Florida take no action to prohibit the teaching of evolution in the schools led the Florida Educational Association at its Tallahassee convention to pass a resolution expressing belief in the wisdom of perpetual separation of church and state affairs." The association's resolution was carefully worded: "The teaching of evolution may be improper, but the study of evolution is proper, in the schools. To explain this choice of words, we mean that it may be wrong for the state to have books and teachers teaching the pupils that a certain definite theory of evolution is a fixed fact. But it is right that the various evolution findings of distinguished scientists be studied by the pupils for whatever they are worth, in the furtherance of knowledge."[29]

The resolution was to no avail, though.

A State of Chaos

Leo Stalnaker certainly didn't like it that dance halls operated on Sundays. When the young lawyer ran for a seat in the Florida House of Representatives in 1926 he was keenly focused on the betterment of public morals. He wrapped up his campaign speeches with dramatic poetry readings asking God to guide morally upright men into positions of

Judge Leo Stalnaker (*center*) in Tampa, 1927. (Courtesy Tampa-Hillsborough County Public Library System.)

leadership. He defended Prohibition, consequently making a name for himself as a real law-and-order man. "Half-way methods do not meet his approval," a newspaper article explained.[30]

Stalnaker won a Hillsborough County seat in the House at the age of twenty-eight. He quickly set to work drafting bills outlawing the operation of dance halls on the Sabbath and requiring all schools to fly the American flag when in session. Stalnaker also wanted to give police the authority to search vehicles for contraband alcohol with or without a warrant.[31] However, those issues were trivial compared to his signature piece:

> House Bill No. 87:
> A bill to be entitled An Act to Prohibit the Teaching in any School, supported in whole or in part by Public Funds, of any Theory that mankind either ascended or descended from a lower order of animals, or any theory of Evolution pertaining to the origin, ascent or descent of mankind, or any theory not in harmony with the Biblical account of the creation of mankind, and to prohibit the use or adoption for use in any school, supported in whole or in part by public funds, of any textbook which contains any theory that mankind either ascended or descended from a lower order of animals, or any theory of evolution pertaining to the origin, ascent or descent of mankind, or any theory not in harmony with the Biblical account of the creation of mankind, and fixing the penalty for the violation of this Act.[32]

There was quite a buzz when the bill was officially introduced. Legislators were generally supportive, but for a wide variety of reasons. For example, Speaker of the House Fred Davis revealed that he wasn't crazy about the bill but that he couldn't "vote against a bill of this kind, because my people would approve of it."[33] W. T. Hendry and L. J. McCall only wanted restrictions on the teaching of the evolution of man. B. M. Frisbee fully supported the bill, saying that teaching evolution violated the separation of church and state. On the other hand, the majority of newspaper editorials blasted the bill, calling it a waste of time that would make the state a laughingstock.

State colleges had to be careful in how they reacted to the bill. Their budgets depended upon the goodwill of the legislature, and Stalnaker had made it clear that he wasn't above wielding that stick if he had to. However, Dr. Hamilton Holt, the president of private Rollins College, didn't have to worry about such things. He told the *New York Times*, "Florida should not become the butt of ridicule throughout the country as has the state of Tennessee by the passage of the anti-evolution bill now before the legislature." The college's newspaper, the *Rollins Sandspur*, was proud to proclaim in its May 6, 1927, issue that Rollins was the first college in the state to take any action in the matter. Dr. Holt wrote in that issue, "Last year I declined a substantial gift for Rollins College, which was proffered on the condition that we should never teach the theory of evolution. I replied to the would-be donor that no self-respecting college would accept a gift under any conditions that limited its freedom of teaching."[34]

The anti-evolution bill's first stop was an education committee, of which Stalnaker was a member. The committee hosted a public hearing on April 19 that "attracted an audience that taxed to capacity the spacious hall and adjoining anterooms of the House, many persons having attended from neighboring towns in this section and South Georgia." At the hearing, Stalnaker was fired up, claiming that textbooks in the state's high schools and colleges contained things "too vulgar to be mentioned before a mixed audience." Speakers railed against atheists and agnostics and blamed a rise in campus suicides on students' exposure to evolution in the classroom. "The taxpayers have a voice in what our children are taught," Stalnaker said. "I am not willing to have any questionable and unfounded theory expounded for them. If we're to raise up a race of Godless infidels, we have no business here. Evolution is a danger at our door."[35]

There was opposition to the bill, however. Rep. F. L. D. Carr from Hillsborough County called it "foolish, needless legislation which can serve no good purpose." Carr was one of the bill's most vocal opponents as the legislative session wore on, and he warned from the outset: "Think of the reputation Tennessee earned for herself by her monkey measure."[36] However, the education committee rolled right over such voices and passed the bill unanimously. Without any other committee stops to worry about, the bill just needed to be put on the calendar for the full House to consider.

At least one lawmaker couldn't wait for the bill to hit the House floor, though. Near the end of April, when representatives were working on a bill that would create the positions of two assistant supervisors of elementary schools, Rep. A. W. Weeks from Holmes County attempted to amend it. He wanted to require that the new supervisors determine "whether any teacher is engaged in the teaching of atheism or agnosticism or is teaching as true Darwinism or any other hypothesis that links man in blood relationship to any other form of life, or is teaching any principle or precept that is in derogation of the teaching of the Holy Bible."[37] If a teacher was found guilty, Weeks wanted a report about the offense printed in a "conspicuous place" in a local community newspaper. Additionally, the teacher would "be deemed guilty of a misdemeanor, and fined not exceeding ten dollars and imprisoned not exceeding five days, and shall thereafter be forever disqualified to teach in any public school in the State of Florida." Weeks's amendment was shot down. Even Stalnaker opposed it, assuring everyone that his bill would satisfactorily address the evolution issue when it came up for its second reading.

Meanwhile, organizations throughout the state began to choose sides. Attendees of the Annual Conference of High School Principals in April issued a defiant statement that "only that education can be free which allows complete liberty to seek the answer to any honest question and we would respectfully remind our fellow citizens that while legislation seeking to control the subject matter of the curriculum may impede educational progress it has not the power to alter, modify or set aside any immutable law of nature, science or God." The Florida Medical Association adopted a resolution opposing the bill, reasoning that "the passage of such a bill into a law would handicap the advancement of medicine and other sciences in the state of Florida."[38]

The controversy also sparked the creation of the Florida Society for the Promotion of Scientific Knowledge, headquartered in Tampa. A meeting in late April featured several speakers who certainly didn't pull any punches, such as one who called anti-evolutionists "bigots who do not think at all." An advertisement in the *Tampa Tribune* countered this new society by asking anti-evolutionists to come together for a meeting to "Stop this Monkey Business in Our Schools." About two hundred people attended to hear speakers protest against the "evils" of evolution. They claimed that "exponents of Darwinism were threatening to destroy the principles of the teachings of Christianity."[39]

As the furor escalated outside the capital, legislators prepared for battles of their own. Despite its easy passage through the education committee, the bill slammed into a roadblock on the House floor. On May 6 the bill unexpectedly "burst like a bombshell" when opponents had purposely sped through other business, including postponing discussion of Stalnaker's bill forbidding dance halls to operate on Sundays, in order to clear way for it. The goal of the surprise tactic was to force a vote on having the anti-evolution bill booted down the calendar to the very end of the legislative session. Stalnaker vigorously fought the move, and after a "spirited skirmish by proponents and opponents" his bill instead secured a special order spot on May 11.[40]

Anticipating a dramatic show, spectators jammed the House balcony a half hour before the event, and many had to be turned away for lack of room. Those who managed to get inside weren't disappointed. First, Stalnaker proposed an amendment that would allow voters to decide the fate of his bill rather than lawmakers. "It remains to be seen whether the lives of our children should be blighted by the teaching of evolution," he said. This was followed by a flurry of other amendments that watered down the bill language. Instead of forbidding the teaching of evolution outright, the proposed alternatives sought to affirm that no teachings would deny God's existence or run counter to the Bible. Their stated purpose was to "clarify the bill and prevent any possible legal conflict on its constitutionality."[41]

As the amendments piled up, it became evident that many lawmakers were concerned about the bill's legality. Taking advantage of the confusion, Carr made a successful motion to send the whole mess to a joint judicial committee for further review. Supporters of the bill saw this as a stalling maneuver and attempted to minimize the damage. The judiciary committee was typically allowed six days to report back to the full House, so a motion was made to cut that down to just two days. After a brief fight, the motion was narrowly approved.

The combined five judiciary committees debated the bill and its amendments that afternoon. There were plenty of anti-evolutionists represented, but the wide variety of wording choices split them. The end result was that the bill was killed on a 19–17 vote. For most bills this would have been the end of the road. However, that evening supporters of the anti-evolution bill held a public rally at a local high school auditorium. Washburn was the master of ceremonies. Legislators gave speeches about the merits of the anti-evolution bill, and even former Florida governor Sidney Catts, who was hoping to make another run for the office, offered his stamp of approval. The crowd wasn't as large as the organizers had hoped for, but it was an energized event nonetheless.[42]

"Evolution Phoenix Plunges House into State of Chaos," a *St. Petersburg Times* headline announced a couple days later. Bill supporters essentially halted all other business on Friday, May 13, in an attempt to resurrect the bill. If they could round up two-thirds of the legislators to help them, they could override the judiciary committee's report. Likewise, if opponents could muster enough support, they could permanently bury the bill. Unfortunately, neither side could reach its goal, so endless debate and stalling tactics propelled the session past midnight. "With a dozen members clamoring for the floor, the speaker adjourned the body without action, and ordered the chamber cleared," the *Tampa Morning Tribune* reported.[43]

But Weeks ignored the order and took over the podium. "Rapping constantly for order during this noisy aftermath," he assembled an unofficial committee tasked with writing an acceptable compromise bill. "We won't allow any bill to pass this House until this anti-evolution bill goes through," he said. "If they want a filibuster, by the eternal Gods we'll give it to them."[44] However, a "gentlemen's agreement" was eventually reached that didn't necessitate yet another new bill draft. Lawmakers would vote on May 17—without resorting to filibusters—on one of the previous, watered-down amendments that had been offered before everything was killed by the joint judicial committee. Both sides realized that too much time had already been wasted and that many other bills needed attention. After a brief bit of haggling, a 67–24 vote approved a final bill that didn't directly mention evolution:

> It shall be unlawful to teach as fact in any school supported in whole or in part by public funds in this state any theory that denies the existence of God, that denies the Divine creation of man, or to teach in any way atheism or infidelity.

It shall be unlawful for any professor, teacher, instructor or text-book committee or commission to use or adopt for use in any school in this state, supported in whole or in part by public funds any textbook which teaches as fact any theory that denies the Divine creation of man, or which teaches atheism or infidelity, or that contains vulgar, obscene or indecent matter.

Violation shall be punishable by a fine not over $100.[45]

The days spent battling over this bill had been exhausting and frustrating. Legislators left ample evidence of this in the official House record. Carr chose to express his thoughts in poetry:

I am now a legislator. Ah, woe to me!
I'm between the Devil and the deep blue sea.
This bloody evolution has already "got my goat."
On the blasted, bloomin' question I don't know how to vote.
To gain my next election, I know the bill must pass,
So I guess I'll ape the monkey by voting like an ass.[46]

Reps. Fuller Warren and R. E. Oliver wanted Carr declared "the poet laureate on evolution of the State of Florida; and since the said Major Carr is between the devil and the deep blue sea, that we extend our sympathy to him in case he goes either place."[47] Representatives jokingly agreed to have the resolution referred to the Committee on Live Stock.

Carr then offered House Resolution No. 60:

WHEREAS, The House of Representatives has for the past fifty days had inflicted upon its membership sudden and unwarranted outbursts of oratory; and

WHEREAS, Said infliction of tongue-torturing tactics has accomplished no purpose other than killing many meritorious measures; and

WHEREAS, The efforts of the effulgent oratory opponents has resolved into a contest of speed and endurance between Hon. A. W. Weeks and Hon. Fuller Warren, of the "Amen" side of the House, and Hon. J. M. Lee and Hon. W. A. Mackenzie of the "Evolution" side; and

WHEREAS, The fore-mentioned gentlemen never having been

given unbridled liberty to extol their spontaneous outbursts; therefore be it

RESOLVED, That during the "Third House" session, on the night of May 28th, these gentlemen each be allowed such time as they individually require, to spout and shout on the following subject, and which appears to be their individual favorite topic: "What I Have Accomplished in This Session of the Legislature"; and be it further RESOLVED, That the members of the House shall judge as to the merits of the tongue exercise and award to the winner of a leather medal, suitably engraved, setting forth the fact that such winner is qualified to return to the 1929 session of the Legislature and be worthy of employment in the capacity of reading clerk.[48]

A few days later the anti-evolution bill appeared in the Senate, but lawmakers there, unlike their House counterparts, refused to waste time on it. A *St. Petersburg Times* article reported: "When introduced in the Senate, the bill immediately became a bauble, to be tossed back and forth between committees until it finally found a resting place with no report."[49]

That didn't mean the bill had no supporters. Sen. John S. Taylor from Pinellas County, who was also vice president of Bible Crusaders of America, tried to have the bill moved to the Senate floor without a committee review. He wanted to sneak it out there by referring to it only by bill number rather than by its title, hoping no one would notice the maneuver.[50] Education committee members caught on, though, and stopped the bill in its tracks. The session was nearly over and there were many other more important bills to juggle, leaving the anti-evolution bill out in the cold. The bill was finally dead, this time for good.

As with any good tale of suspense, though, there was a plot twist. The whole business about school textbooks being full of "vulgar" material had caught lawmakers' attention. This prompted the easy passage of the Senate Concurrent Resolution No. 28:

WHEREAS, It has been brought to the attention of this Legislature that there are certain text books and reference books in use in the institutions of higher learning of the State which are inimical to good morals and healthy thought; and

WHEREAS, It would be for the best interests of the youth of the State that these books be eliminated; therefore, be it

Resolved by the Senate, the House of Representatives concurring, that a special committee of two members from the Senate and two from the House be appointed by the President of the Senate and the Speaker of the House, respectively, whose duty it shall be to work in conjunction with the Board of Control and make a thorough canvass and investigation of all text and reference books used in said institutions, and to report as soon as possible to the State Board of Education any and all books which in their opinion are detrimental to good morals and clean thinking, giving in such report the reasons why such books should be condemned. That upon such report coming in it shall be the duty of the State Board of Education to have withdrawn and eliminated all books which, in their judgment, are objectionable.

RESOLVED, further, That it shall be the duty of the Board of Control to work in conjunction with said special committee to the extent that the purpose of this resolution be carried out.[51]

Sen. W. J. Singletary from Jackson County took the lead in promoting the resolution, telling his fellow senators that shelves needed to be wiped clear "of those stinking books." An editorial in the *Tampa Morning Tribune* lambasted the resolution, saying that few lawmakers "are in possession of any degrees, or experience in the field of education which would equip them to teach a grade school, much less act as inquisitors and censors of text or reference books." This infuriated Singletary, who then barred that paper's editor from the Senate chamber.[52]

It turned out that the legislature was actually a latecomer to the book-banning effort. A citizen group had been on the hunt since the previous year. L. A. Tatum, an elder in Tallahassee's First Baptist Church, had created the Florida Purity League in 1926.[53] The group's main target was the Florida State College for Women and the supposedly heretical books found in its libraries and classrooms.

Tatum had gotten the idea for the Purity League when a disgruntled professor who faced losing his job at the college fed Tatum information about books used by his fellow instructors. The texts were mainly from psychology and sociology courses, and the offending passages from the books dealt with touchy subjects like sexuality or were "pro-German." But going along with the anti-evolution fervor of the time, Tatum didn't leave it out of his group's attacks. To prove just how horrible these books were, one of the quotes he pointed out to the Board of Control—the entity that oversaw the state's colleges at that time—was: "psychoanalysis represents by an extension of the theory of evolution, an application of the principle of evolution to the study of the mind."[54]

Tatum's complaints were heard by the Board of Control over and over again, but they didn't gain much traction. The president of Florida State College for Women, Edward Conradi, defended his school and professors before the board, saying that no one taught anything contrary to the Bible and that the books currently in use in his school were widely used in colleges throughout the country. But then came the 1927 Senate Concurrent Resolution.

Singletary led the legislature's investigative committee. He took the same road Tatum was on, challenging psychology and sociology texts and accusing professors of teaching evolution as fact. This also placed the University of Florida under the gun. University president Albert A. Murphree stood alongside Conradi in defense of their schools. However, Murphree tended to be more conciliatory than Conradi in the beginning. Murphree offered to move all questionable books to a secured area so that only students with special permission and their teachers could access them.[55] Conradi resisted doing so, especially when it came to having to appease Tatum, but he eventually followed suit. He even assured everyone that "if any teacher is found teaching anything contrary to the word of God and the Christian religion, such teachers will be reported to the board of control, and the president will recommend his or her removal."[56]

Conradi's assurance satisfied the committee, and its investigation fizzled out over time. Tatum, on the other hand, demanded much more. He wanted the books off campuses completely, and there needed to be a purging of "dangerous teachers," too! However, his efforts grew more fanatical over time, as his actions, such as spreading false rumors and endlessly adding to his list of books to be banned, grew into an embarrassment. An example of how tiresome he became comes from the Tallahassee Chamber of Commerce: "We are not in sympathy with your propaganda against the Florida State College for Women, Dr. Edward

Conradi and the faculty of the college. It is our opinion that your propaganda is distasteful to the vast majority."[57]

Tatum then found himself the subject of a grand jury indictment charging him and coauthor A. Pichard with sending obscene literature through the mail. They had published a pamphlet titled *Psychoanalysis of "Filthy Dreamers": And Other Insidious Teaching under the Guise of "Science" in Tax Supported Institutions of Learning,* which contained quotes from objectionable textbooks. "We are quite willing to be the victims of persecution and prosecution hoping thereby the true situation in our colleges and universities may be brought home to the minds and hearts of Christian parents," they told the grand jury.[58] The case was later dismissed.

"Disgraceful Even to African Cannibals"

The red-hot anti-evolution effort of the 1920s had cooled off significantly by the end of the decade. It became a minor issue during the 1928 gubernatorial campaign, with three out of the four candidates advocating anti-evolution. Among Sidney Catts's platform issues was: "Let's have no evolution of man in our schools." A broadside distributed on behalf of his campaign proclaimed that Catts "would make an attempt to eliminate this FREE LOVE and MONKEY TEACHING from the Girls' College, which has proven a disgrace to the State of Florida and would prove disgraceful even to African cannibals."[59] One of Catts's rivals was John Taylor, the senator who had tried to sneak the 1927 anti-evolution bill out of committee and onto the Senate floor for approval. Possibly as a sign of the changing times, the only candidate not on the anti-evolution bandwagon, Doyle Carlton, won the governor's office.

Anti-evolution popped up briefly during the 1929 legislative session when the Senate considered a bill about textbook selection. Sen. Joseph Scales wanted to amend it with a provision that teachers could choose not to use textbooks that promoted evolution "contrary to Biblical teachings." Singletary spoke in favor of the bill: "If our children have to be taught things that make us blush to read, it is high time that we incorporate in our laws what this amendment proposes." The amendment drew a hostile response, though. Sen. Pat Whitaker expressed what most other lawmakers were thinking: "This legislature was pestered during its last session by 'monkey business' which emanated from Hillsborough county. I think it's time that we pass a uniform text book law without confusing it with the evolution issue."[60] The amendment was shot down, and the anti-evolution movement overall faded with it. An anti-evolution bill was introduced in the 1933 legislative session, but it didn't make it out of a judiciary committee and attracted no significant interest in the media.[61]

2

"Un-American, Atheistic, Subversive, and Communistic"

It would take three decades for another significant anti-evolution movement to appear in Florida. In the meantime, teachers still had occasional uncomfortable encounters with parents and students when science subjects clashed with religious beliefs. In 1948 the Florida Department of Education published *A Brief Guide to the Teaching of Science in the Secondary Schools*. Written by science educators from across the state during a University of Florida science education workshop, the 128-page book included a section titled "The Book of Genesis, and Science" with the following introduction: "Many teachers asked the workshop group which prepared this bulletin to include suggestions, if any could be found, for helping the teacher advise with students and parents about the supposed conflict between the teachings of science and the teachings of the Bible. Many teachers have been misunderstood in their communities because of the inability to solve this problem. It is a real problem for many students and parents. The requests for help were urgent."[1]

The chapter walked teachers through suggested responses to a hypothetical student questioning the "conflict between interpretations of the text books of science and the Book of Genesis in the Old Testament of The Bible." The main points the book offered were that the Bible should not be taken literally and that science was a method of exploring God's creation. "Science does not deny God, but it attempts to explain His handiwork," the text stated. "If an individual scientist denies God, or an individual theologian denies science, then neither uses his gift of high understanding. True scholars in both theology and science have no turmoil in their thinking."[2]

In 1962 the state department of education issued another publication that also touched on the intersection of religion and science education. The national concept of a God-fearing America pitted against godless communism was flourishing due to prolonged Cold War political and military tensions with the Soviet Union. Some people believed that God and morals were losing ground in the public schools and being replaced with secular teachings that conflicted with family values. In Florida this led the Florida Board of Education to unanimously approve the publication of *A Guide: Teaching Moral and Spiritual Values in Florida Schools*. Thomas D. Bailey, the state's superintendent of public instruction, expressed those sentiments in his foreword: "To be silent about religion and the contributions of God-centered religious thought to the growth and development of our nation may be, in effect, to make the public schools an anti-religious factor in the community. . . . The purpose of this Guide is to assist teachers to develop in students appreciation and understanding of the moral and spiritual values in American democracy and to present more effectively the contributions of religion to our American Way of Life."[3]

Chapters expounded on subjects such as American heritage, basic American beliefs, and the development of moral and spiritual values. There were suggestions on how to inject such values in language arts, social studies, and music classrooms. Science and mathematics were deemed especially important in this endeavor.

> While the areas of social studies and English lend themselves beautifully to the development of topics or units dealing directly with moral and spiritual values, it is believed that foundations for the moral and spiritual can best be laid in the science class through an attitude of the teacher friendly to the idea that in America moral and spiritual values have their roots in God.[4]

Although evolution was never directly mentioned, it was certainly hinted at:

> Hardly any teacher is so blessed by interest on the part of the student or so hampered by preconceived notions as is the science teacher trying to deal with the topics relative to God's part in the creation and development of man and the universe.[5]

The science teacher was singled out as the key to success. It could all come down to a simple "well-chosen word here or there" that would either imply or deny a connection between scientific studies and "the existence of a Supreme Being." There were stern words for those who expressed too strong of an opinion to students:

> Most of the difficulties resulting from discussion of conflict between the Bible and science come about through the willingness of some teachers to discuss a topic about which they know little or nothing. The same teacher who feels that years of study are required to make a good science teacher will gladly discuss Biblical criticism, a subject to which he has given no serious study, at the drop of a hat.[6]

These guides were the only evidence that there were any conflicts over what had been taught in Florida's biology classrooms since the 1930s. It was all quiet otherwise, primarily because there wasn't much to get worked up about. Evolution was essentially out of the textbooks, and for decades there was no motivation to put it back in. There were discussions of general evolutionary concepts in many books, but the word "evolution" was typically avoided and wasn't a main theme.

In the mid- to late 1950s, America's growing rivalry with the Soviet Union challenged the education system's status quo. There was heated criticism that science and technology education at all levels had lost ground to the Soviets and desperately needed revitalization. In response, the U.S. government pushed educators to overhaul their stagnant, rote-memorization-of-facts style of science instruction. Then the Soviets launched the world's first artificial satellite, Sputnik, into Earth's orbit in 1957. After that confidence-shattering event, the race against the Soviets for superiority became a national priority. Congress quickly responded with the National Defense Education Act in 1958, which, in part, infused states with cash to develop better practices and materials for teaching science, mathematics, and foreign languages.

All branches of science and mathematics education were under scrutiny, but biology had a special burden to bear. The vast majority of high school students took a biology course, far more than took courses in chemistry and physics. This meant that many young men and women's only significant exposure to science was their high school biology class. With this in mind, the Biological Sciences Curriculum Study (BSCS) was created by the American Institute of Biological Sciences using a grant from the National Science Foundation. Its task was to draft brand-new biology textbooks and reform the way biology was taught.

BSCS's first director was Arnold Grobman. At the time he was offered the position, Grobman was director of the Florida State Museum (later renamed the Florida Museum of Natural History). He initially turned down the offer because he was happy with his Florida job. Grobman got an earful from his wife, though: "You're a damn fool!" With his wife's vocal support, Grobman moved to BSCS's headquarters in Colorado, where he was joined by a cadre of enthusiastic professionals from across the country. "I was actually involved, on almost a daily basis, with some of the nation's top biologists and outstanding high school teachers in a highly charged intellectual atmosphere," said the herpetologist.[7]

The materials they produced over the next few years were tested in school districts in nearly every state. In Florida, ten Dade County biology teachers—two from each senior high school—participated in BSCS workshops and seminars throughout the 1961–62 school year. Their school board members were enthusiastic supporters, authorizing the purchase of $21,134 worth of the new BSCS books and materials in a unanimous vote.[8]

"I Wish the Lady Luck"

However, that eager, positive attitude was tested in October 1962 when an outraged parent complained during a Dade County School Board meeting that the biology books were "un-American, atheistic, subversive and Communistic." Rev. David Brandt Berg was in charge of a local branch of Soul Clinic, an evangelical missionary and "witness training" school. His son returned home from his Miami high school one day upset about the predominance of evolution in biology class. "He complained to me and when I read the book, I was shocked," Berg said. "I took him out of school immediately."[9]

Berg spoke again at the school board's November meeting, claiming that use of the new textbooks was illegal and irreligious. Board member Eunice Anderson disagreed. The meeting's minutes reported that "in her opinion a person's faith would have to be rather shaky if a scientific approach such as this could shatter his confidence." The board dismissed Berg's protests. Feeling he had no other option, Berg filed a complaint with the state attorney's office, "charging the Dade School Board violated state law and constitutional rights by permitting use of a biology text about evolution." He claimed to have a petition signed by thousands of supporters from across the country. "I even have been approached by several biologists willing to testify on my behalf," he said. State attorney Richard Gerstein turned away Berg's attempts to ban the books. He couldn't see how the texts broke Florida law, and he didn't find anything "objectionable" in them.[10]

As news about Berg's protests got out, several people came to evolution's defense. "The Florida Academy of Sciences unreservedly endorses the use of the BSCS High School Biology, Blue Version text, in the teaching of Biology courses," wrote Dr. Alfred P. Mills, the academy's president, in a letter to the school board. "The treatment of evolution in this text, as a proper subject of study, and as a unifying biological concept, is in accord with the thinking of present day biologists." Additionally, a Unitarian Church pastor, a representative of the Greater Miami Council of Churches, and a delegate from the American Humanist Association all told school board members that they supported evolution's role in the new biology courses.[11]

Berg's wife, Jane, tried a different approach in January 1963. She gave school board members copies of the book *Why We Believe in Creation— Not in Evolution* by Fred John Meldau. She hoped they would consider including it in the biology curriculum "and that the children be given the privilege of choosing the one they want to study."[12] The board declined to comment on the proposal.

The crusade soldiered on in February when Berg announced that he was going to take one of his son's former teachers, Elizabeth Ferguson, to court. "She will be our test case, because my 15-year-old son, Paul, was one of her pupils and can testify to what she taught." Berg was confident that he had a strong case. "Apart from anything else, Mrs. Ferguson is teaching from a book not adopted by the State Legislature as required by law." Ferguson explained that the textbook was brand new and had not yet been officially adopted, and she assured reporters that the potential lawsuit didn't upset her. "I insist this book teaches evolution only as a theory," she said. "I am a Roman Catholic and I would not teach it if it were an affront to any religion."[13]

Newspapers set up the escalating conflict as the next "Monkey Trial." A *Miami News* headline announced: "Repeat Of Famed Scopes Drama Scheduled Here." Reporters even asked sixty-two-year-old John Scopes—who was living in Louisiana then—for his thoughts on a future Florida evolution trial. "I wish the lady luck," he said of Ferguson. Scopes also affirmed his support for the teaching of evolution: "I have never changed my mind over the years."[14]

The highly anticipated trial never materialized, because Berg couldn't find an attorney willing to take his case. He then moved to Texas after claiming that city and county officials had condemned his Miami home in retaliation for his agitations against evolution at school board meetings and his outspoken religious beliefs. In the late 1960s Berg moved to California and founded a new religious movement called Children of God, which later attracted extensive publicity for its controversial views and practices.

"Grievously Offensive and Inexcusably Reprehensible"

It didn't take long for another striking personality to rise up to the challenge of fighting evolution's resurgence in Florida's schools. Rev. Clarence E. Winslow was a retired minister from the First Church of the Nazarene in Clearwater and a chaplain of the Kenneth City Police Department. The sixty-four-year-old had been doing a lot of reading in his free time, and he was kicked into action by a 1971 newspaper report on the newest anthropological research. The article overall upset him, but what really fired him up was this quote: "Dr. Russell H. Tuttle, a University of Chicago anthropologist, subscribed to the theory that man's forebears were a special species of apes which began walking upright soon after leaving the trees."[15] The article went on to say that some scientists thought humans never went through a "knuckle walking" phase, while others said they did. All of that was offensive nonsense, Winslow thought. He wrote a letter to the Pinellas County School Board, citing that news article and following it up by saying, "The Genesis record indicates that God created Adam and Eve as adults and pronounced them husband and wife."[16] Thus began Clarence Winslow's long campaign against the teaching of evolution in Florida's schools.

Winslow wrote an article he titled "The Bible and Evolution—The Winslow Resolution." It stated that the Bible had been abolished from public schools by atheists and "certain others" and that children were being exposed to the theory of evolution, which "is grievously offensive and inexcusably reprehensible to Bible believers."[17] He called for action. He wanted people to sign his resolution and "correct this un-American imbalance in our Educational process." He found his first sympathetic audience at the Pinellas County School Board.

On August 25, 1971, Winslow gave a presentation to the school board members. They liked what they heard and accepted the "Winslow Resolution" in a unanimous vote.[18] This didn't require that the school district actually do anything, though. The board simply added its authority to the resolution so that Winslow could proceed with his quest. He wasn't worried about what was happening on the local level; he had set his sights on much bigger goals. He wanted the state legislature to create a law that would require schools to teach not only evolution but also the biblical account of creation. After that, he wanted to solicit the U.S. Congress.

Propelled by the Pinellas County School Board's approval, he moved on to the Pinellas Legislative Delegation and successfully captured the attention of Rep. Dennis McDonald from St. Petersburg. On February 1, 1972, McDonald filed a bill summarized as:

HB 2937—A bill to be entitled An act relating to education; directing the school board to require in the teaching of evolution the reading of appropriate religious passages dealing with creation and evolution including, but not limited to, chapters one and two of Genesis; providing an effective date.[19]

In an article announcing the filing of the bill, Winslow wrote: "It is utterly unfair to teach young people the unprovable theory of evolution without including . . . the biblical account. Kids are growing up not even knowing what the Bible is."[20]

Representative McDonald was aware that his bill would be heavily challenged, especially on the grounds of separation of church and state, but he wasn't deterred. "It's not denominational in any way, shape or form," he said. "It's one of two widely accepted theories. To drop it would be the equivalent of dropping any reference to biblical history."[21] The bill was referred to the House Committee on Education but failed to gain any traction. After the initial article in the news announcing its filing, the media didn't mention it again. Although the issue publicly died out for a year or two, Winslow remained busy behind the scenes researching and planning his next move.

"Men Have Always Been Men"

A lull in the Florida-based creationists' activities was offset by nationally known creationists making inroads into the state. In March 1974, two prominent representatives from the Institute for Creation Research in California, Dr. Henry Morris and Dr. Duane Gish, accepted an invitation to a debate in Tampa. Morris had founded the institute in 1972 to research and promote the teaching of biblical creationism with an emphasis on public-outreach efforts. The pair took on two professors from the University of South Florida, anthropologist Evelyn S. Kessler and biologist John V. Betz. The well-known locals took the affirmative stance on the debate resolution "Evolution is a better model than creation with which to explain the scientific data related to origins."[22]

The event was well publicized, and about eight hundred eager spectators packed into Tampa's Plant High School auditorium. Despite being the local representatives, though, Kessler and Betz found themselves in hostile territory. A newspaper exit poll showed that about four hundred audience members considered themselves creationists, while only ten people supported evolution. This was due to Christian organizations busing in supporters from out of town, many of them teens from Christian youth camps.[23]

Gish and Morris said they were not there to confirm or deny the Bible. They wanted to show that "the scientific and historical data available" clearly indicated direct acts of creation, not evolution: "That bats have always been bats, cats have always been cats, and men have always been men," said Gish. Kessler tried to define and describe evolution for the audience, saying that humans didn't descend from apes but rather that apes and humans descended from a common ancestor. "The evolutionary change has been slow, almost imperceptible over millions of years," she said. Gish countered that there was a lack of transitional forms in the fossil record. "Each form in the fossil record appears abruptly, even the earliest form," he said.[24]

Betz admitted that there were gaps in the fossil record but said that they were to be expected, just like when a person discovers gaps in a personal family history: "He knows he has a grandfather, and even if there is no proof, he assumes his grandfather had one too." Betz went on to say that he believed God created everything, but not as depicted in the Bible. Evolution "is a better, more natural, more coherent explanation" than religion for the world seen today. As the debate wrapped up, Gish and Morris said they believed creationism should be taught in the schools. "For three generations, evolution has been taught exclusively. And that's unrealistic in a free society," said Morris.[25]

A couple of weeks after the Tampa debate, another debate was held in Fort Lauderdale. This time, though, neither of the featured debaters was from Florida. Instead, Gish took the stage at Coral Ridge Presbyterian Church against Dr. Joel Warren from Nova University. The event organizer, Dr. D. James Kennedy, the megachurch's founder and the creator of the Coral Ridge Ministries national broadcast organization, said that he had called several Florida universities looking for someone to challenge Gish but couldn't find a single interested person. "Dr. Kennedy was amazed at the reluctance of these professors, who teach evolution to their students as an established fact year in and year out, to publicly face a challenge from creationists on the comparative scientific validity of the creation and evolution models."[26]

Scientific Creationism in Florida

In August 1975, University of Texas professor Harold Slusher visited St. Petersburg to give a talk called "A Look at Creationism versus Evolution."[27] Slusher's biography said that he was a research associate in geoscience for the Institute for Creation Research. He had also coedited the high school textbook *Biology: A Search for Order in Complexity*. The talk was sponsored by a local organization called the Committee for Creationism in Education, headed by Rev. Clarence Winslow.

Winslow had learned a thing or two about how to challenge evolution's place in the schools when his push for legislation had failed. He consulted with the state commissioner of education and the state chief of curriculum.[28] The revealed problem was that the Bible simply wasn't ever going to get into the science classroom legally, and attempts to show the scientific validity for evolution opposition lacked educational materials to support it. But when Winslow discovered *Order in Complexity*, his prayers were answered.

Order in Complexity was developed by the Creation Research Society, an organization formed in the early 1960s to conduct research and publish materials advocating "special creation"—based on a literal interpretation of the Bible—as an alternative to evolution. The organization was incorporated as a nonprofit in Michigan, but it boasted a robust international membership, including Slusher. The book was published in 1970 in response to the renewed emphasis on bolstering science education in the United States after the Soviet Union had launched Sputnik. Biology textbooks produced at this time emphasized evolution, and Creation Research Society members felt they needed to counter those texts with one of their own.

Order in Complexity stated in its preface that "the most reasonable explanation for the actual facts of biology as they are known scientifically is that of Biblical creationism." An entire unit critiqued evolution as lacking supporting fossil evidence, relying on inaccurate fossil dating methods, and claiming that gene mutations can be beneficial. The authors hoped private schools would consider *Order in Complexity* as a serious biology textbook alternative. "We trust it will also be of interest to public school systems desiring to develop a genuine scientific attitude in their students, rather than an artificially-induced evolutionary world-view," they said in the preface.[29]

A few days after Slusher's talk in St. Petersburg, Winslow was back in front of the Pinellas County School Board, toting along with him one hundred copies of Slusher's textbook. Winslow donated the books to the school system, saying: "The psychological impact of the theory of evolution has reduced man's image of himself from a divinely created personality to a mere animal. We are witnessing the surfacing of animalism." Since the Slusher text was not on the official state-adopted textbook list, Superintendent Gus Sakkis said it would be made available to biology teachers as a resource book only. Upon hearing this, Winslow asked for the board's help in getting it on the state list. Whether or not they responded to this request is not recorded, but it's worth noting that school board member Ron Fisher also happened to be the vice chairman of the Committee for Creationism in Education.[30]

During his presentation to the school board, Winslow mentioned that he was also discussing the issue of creation versus evolution with school boards in Manatee and Hillsborough Counties. His new strategy was to establish a network of people friendly to his cause on the local level. For instance, he approached the Manatee County School Board in October 1976, introducing himself as the president of the American Basics Council. He requested that the school board consider a resolution he offered about "the teaching of both the belief of divine creation and also the evolution of man."[31] The board approved it unanimously. But just like the 1971 resolution accepted by Pinellas County, Manatee County apparently wasn't expected to take action.

Winslow returned to Manatee County in May 1979. This time he found support from Superintendent William Bashaw, who agreed that theories of both evolution and creation should be included in biology classes.[32] Again, no action was taken by the county school board, but more people were boarding the anti-evolution train. Winslow was making progress.

Order in Complexity never gained much traction in Florida, but it had been an officially approved textbook in other states. Tennessee enacted a state law in 1973 requiring textbooks to provide equal coverage to evolution and biblical creation, with *Order in Complexity* being an acceptable example. However, the law was found unconstitutional in the 1975 federal court case *Daniel v. Waters. Order in Complexity* was also the centerpiece of an Indiana state-level court case in 1977. The judge in *Hendren v. Campbell* ruled that *Order in Complexity* was essentially a creationist book and that creationism was a sectarian religious view. Therefore, using *Order in Complexity* in Indiana public schools was unconstitutional.

These court decisions contributed to the rise of a new concept called "scientific creationism." Anyone wanting to challenge evolution's dominance in the public school curriculum now knew that all references to the Bible needed to be excised. Special creation must be couched in strictly scientific terminology without any ties to specific religious claims. It didn't take long for the new terminology to appear in Florida.

Rep. Tom Bush from Fort Lauderdale and Sen. Joe Carlucci from Jacksonville rode the anti-evolution train to Tallahassee in November 1979. They filed companion bills that introduced the new scientific creationism tactic. A summary described the House bill as follows:

> HB 11-C—A bill to be entitled An act relating to education: creating the "Balanced Treatment for Scientific Creationism and Evolution Act"; providing legislative findings and intent; providing definitions; requiring balanced treatment; prohibiting religious instruction under certain circumstances; providing for nondiscrimination; providing applicability; providing an effective date.[33]

The bill was modeled after a version distributed nationwide by Paul Ellwanger of South Carolina. Ellwanger, a respiratory therapist, sought to combat what he saw as the source of society's ills: the teaching of evolution. The battle lines were clearly drawn. "I view this whole battle as one between God and anti-God forces," he said. However, in a letter to Carlucci he cautioned the senator that it would "be very wise" to keep religious overtones out of any public conversation about the proposed bill.[34] Nonetheless, Bush said in defense of his bill, "The more I study it, the more I find that evolution is much more myth than the account in Genesis could ever be."[35]

But there was opposition to the bill. Rep. Bill Sadowski from Miami said, "It's fascinating that at this day and time something like this could be offered." The bill immediately brought to his mind the Scopes trial in Tennessee, but such things were ancient history, he said. "Those debates are fun and can be meaningful in some times and places, but I don't think they belong on the floor of the Legislature."[36] As was the case with McDonald's bill a couple of years prior, the scientific creationism bill didn't have enough support and died.

The bill's defeat at the state level didn't discourage Winslow. He was a fixture at his local school board meetings, hopping back and forth between Hillsborough County and Pinellas County while also making time for Manatee County. By 1979 he was charging forward at full speed as

Rep. Tom Bush (R-Fort Lauderdale) listens to a debate in the Florida House of Representatives, Tallahassee, May 21, 1981. (Donn Dughi, State Archives of Florida.)

an influential support network grew up around him. He now headed a group called Compatriots for Academic/Religious Freedom. It's unknown if this was just a name change to the Committee for Creationism in Education or the American Basics Council, or if it was a new group altogether.

Winslow found that public opinion was swinging in his favor and that inserting creationism into the public school curriculum was now a real possibility. He just needed to find a school board willing to take the leap, and it looked like Hillsborough County was primed for his final push.

"Supreme Court Approves"

At a December 1979 meeting, the Hillsborough County School Board granted Winslow fifteen minutes to state his case. He came prepared with a stack of papers and delivered an impassioned speech overflowing with confidence and enthusiasm. One document he presented to the board was titled "Reverse Academic/Religious Discrimination Violates Student's Rights."[37] It was Winslow's belief, based on exhaustive research into several U.S. Supreme Court cases, that students were being denied the right to freely express and explore religious belief in tax-supported schools. Children should be allowed to pray and spend equal time on both evolution and creationism.

He chewed up ten minutes of his allotted time reciting a litany of quotes from U.S. Supreme Court justices and Court documents that appeared to support his stance. At one point a school board member interrupted Winslow to let him know that his time was running out. Finally, Winslow got to his point and presented a "Resolution for Creation and Prayer in Public Education—Supreme Court Approves."[38] He wanted them to implement a policy to correct what he saw as a violation of constitutional rights. The resolution also called for textbooks to include creationist concepts and requested that the media "cooperate in clarifying prevailing misconceptions and misinterpretations of court rulings in relevant cases." The retired pastor then demanded to know the status of religious freedom in Hillsborough County schools.

The response was fascinating. The minutes of the board meeting give the impression that Superintendent Raymond Shelton and Assistant Superintendent for Instruction Frank Farmer were giving a full report to Winslow, and there seemed to be a tone of defensiveness. They said that whereas prayer was not mandatory, there was nothing stopping "mature students" from reading material concerning creationism. However, creationism was not part of the curriculum. The pair reminded the school board that Winslow had donated several creationism-based biology textbooks to Hillsborough County in 1975 and that those books were still available throughout the district as reference material. Then Farmer noted that "the vast majority of the science teachers in this school district were church members (primarily of the Protestant faith) and had had the opportunity as science teachers and religiously-oriented people to face this particular problem."[39]

Winslow's final comment of the meeting sniped back at the superintendents. Even though his donation of creationism-based textbooks a few years back was a positive thing, Winslow believed that "board members were misinformed about what is being taught in the schools inasmuch as some students were being taught the theory of evolution and not being allowed to discuss creation on the basis that it is religion."[40] That's a big no-no, claimed Winslow, since his extensive research suggested that the U.S. Supreme Court had decreed that it was not right to teach one without teaching the other. With his resolution before the school board and the conversation now shifting into high gear, Winslow stood back and let the members take it from there.

Board member Roland H. Lewis had an interesting take on where religion and science intersect. The meeting minutes report Lewis's comments: "Some people do not understand that the Bible is a book which deals with history, literature, and science. It is the first published book which taught that the world was round, that the ocean was in a circle of the deep, and that the earth hangs on nothing. Because these scientific facts are found in the Bible does not make the scientific truths become religion." Lewis was in favor of "giving teachers support to teach creation as a scientific explanation as valid as evolution."[41]

Lewis then moved that the board adopt a resolution allowing the teaching of creationism as an alternative in the school district. Overall, his fellow board members agreed that creationism needed to be added to the curriculum, but there were differing opinions on how this should be done. Lewis thought that evolution and creationism should be taught side by side. Sam Rampello agreed that creationism should be taught, but he did not believe it should be presented as an alternative. Rather, he preferred to have the district administration "bring in the theory of creation as one of the methods taught in biology, social science, and other sciences."[42] Joe E. Newsome focused on textbooks, expressing his desire to only purchase those that covered both evolution and creationism.

Farmer then boldly spoke up on behalf of the district's teachers. He asked that the science teachers be given a chance to react to the proposed resolution before the board approved it. The consensus of the members thought that was a good idea, but Lewis had serious reservations. The meeting minutes report: "Member Lewis was willing to abide by decision to defer action on his motion, but did not want 'nothing' to happen. It was his judgment that people who believe evolution through the years had hidden scientific evidence, but the press had followed these things and was reporting the discoveries that are being made now that had been hidden for nearly a hundred years or over. The more discoveries made, the more valid the creation explanation becomes."[43]

A representative from the Classroom Teachers Association, Sam Rosales, spoke up at this point to caution the board against making a decision that it was perhaps not qualified to make, noting that such a decision could run against the grain of academic freedom. Essentially, teachers could be forced to teach something they don't feel they should. This apparently rumpled Rampello's feathers. The meeting minutes report: "Even though Board might not like what the teachers are saying, he [Rampello] thought they should be heard. He hoped that teachers were trained in teaching the theory of evolution as well as creation without indicating their opposition to one or the other. He did not believe certain teachers should be told to choose one or the other."[44]

Chairman Ben H. Hill finally wrapped up the discussion by declaring that action on the topic would be deferred until the next scheduled meeting in two weeks.

"The Failure of the Educational System"

On December 12, Winslow moved on to the Pinellas County School Board with a shorter but no less impassioned demand that students be given their constitutional rights to "freedom of speech and inquiry and of belief." Once again he found a sympathetic audience. Board member Calvin Hunsinger moved to adopt the resolution Winslow gave them. Then a "parade of ministers and a college biology teacher" spoke in support of the resolution.[45] The teacher was Colin Nevin from St. Petersburg Junior College, who told the board that "a growing number of scientists" support scientific creationism more than evolution.

Much like Hillsborough County's school board, the Pinellas County board was all for incorporating creationism into the curriculum but decided to table the motion in order to do further study. There were some nagging questions to consider. For instance, the *St. Petersburg Times* reported:

> "I have no qualms for teaching alternatives," Supt. Gus Sakkis said. But he questioned whether the schools would also have to teach all the theories of creation espoused by the world's religions.

"Indeed, within our own Christian religion, there is some variance in interpretation," Sakkis said.[46]

A few days later, the Hillsborough County School Board faced a challenge as they revisited the creationism subject: a packed house. Before they could even tackle the deferred resolution, they needed to decide how to handle all the people who had shown up to speak. Eventually, it was decided that each person would get seven minutes. Not only were citizens lined up to speak, but the school district administration had been flooded with calls and letters over the previous couple of weeks. They had stacks of papers sorted into pro-creationism and pro-evolution piles. Position papers came in from the American Association for the Advancement of Science, the National Association of Biology Teachers, the National Science Teachers Association, and the American Association of Biology Teachers.

At the direction of the school board, the district staff had written a proposed statement concerning the teaching of creationism and evolution for the members to adopt:

> The Hillsborough County Public Schools will teach as a part of the curriculum in accordance with objectives adopted, evolution as a theory, not as a fact, and will direct teachers to inform students that other theories of the origin of the species based on faith do exist and occupy an important part of our body of knowledge and human thought and are not incompatible with the theories of evolution. Students in the Hillsborough County Public Schools will be encouraged to examine and study these theories based upon faith, utilizing resource materials that are available in all schools.[47]

Superintendent Shelton told the board that he believed in separation of church and state and that taking creationism any further than the proposed statement suggested would invite trouble.

The board members didn't just disagree with the statement, they hated it. They scolded Shelton and his staff for giving them a statement that was far off from what they had asked for. The statement was biased toward evolution, said Lewis. He had serious problems with the use of the word "faith" in association with any theory other than evolution. The meeting minutes reported: "[Lewis] said the issue was not whether it is faith or science; they are both science and it takes as much faith to believe evolution because there is no foundation for it. Accordingly, he felt the statement from the staff was unacceptable because of its being biased." A. Leon Lowry agreed, saying that the statement "totally ignored the other side." Joining the chorus was Marion Rodgers, who said the statement was unacceptable to her. She didn't like how evolution was "focused as being the only thing."[48] Newsome was adamant that either evolution and creationism would be taught side by side or neither would be taught at all.

Shelton weathered the onslaught as best he could with support from the district's supervisor of secondary science, Nancy Marsh. They tried to explain to the board members that evolution had clear scientific support and that the biology curriculum was based on objectives from the State Compendium of Science Objectives, which had gone through a long, arduous approval process. It was also pointed out that teachers usually preface evolution lessons with a statement that the concepts are not meant to interfere with students' religious beliefs in any way.

The board members didn't take kindly to this resistance. Lewis claimed that he had "at least six or seven years of concentrated study in both the theory of evolution and the theory of creation."[49] Hill reread the resolution Winslow had given the board at the previous meeting in an apparent effort to explain to the district staff what they should be working toward. Shelton and Marsh were obviously on the losing end of this fight. Out of seven board members, not a single one went on the record in support of the superintendent's view.

Then the floor was opened to public comment. Of the seven people who spoke, only two supported the teaching of evolution without creationism. One of those two was a student from Plant Senior High School. Unfortunately, his appearance set him up as a target for the other speakers. A college professor claimed that the teen "was totally misinformed," and an attorney said that "the young man from Plant High School was a perfect example of what was happening because of the failure of the educational system to provide an alternative theory which clearly exists."[50]

One point brought up by the pro-creationism speakers was that scientific creationism was "based on hard, cold scientific facts" and had been for ten to fifteen years. Supposedly, the working scientists at universities across the country knew this. When the floor was turned back over to the board members, Rampello was upset. He wanted to know why the colleges weren't training high school teachers about creationism if the professors had been so confident in it for the past decade or more. "If the universities are remiss in that responsibility, he [Rampello] thought they were failing miserably."[51]

It was obvious that the board was eager to correct perceived wrongs and get creationism into the classroom, but it needed an official position statement produced by the school district staff in order to proceed. Hill directed Shelton to take into consideration all that was said during the meeting and produce an acceptable document. The meeting was then adjourned.

Non-scientific Theories

In 1974, Dr. Betz had argued on behalf of evolution during a debate against Dr. Gish, a creationist from California. The two shared a stage of sorts again on March 18, 1980, when they stood before the Hillsborough County School Board to talk about introducing scientific creationism into the schools. Gish said that he was there on behalf of the county's citizens to lend his expertise in explaining the need to correct an imbalance in the teaching of origins.

According to the meeting's minutes, "Dr. Gish made clear that the goal of creationists is not to teach the Bible, the Book of Genesis, or religion in the public schools; creationists do seek to alter the present system whereby most students are indoctrinated or literally brainwashed in the hypothesis and the philosophy of evolutionism." Gish went on to explain that the second law of thermodynamics makes it impossible for a universe to have created itself, rendering concepts such as evolution unscientific and irrational. He claimed that the fossil record is "highly contradictory and embarrassing," and he read quotes from "renowned evolutionists" saying that there are systematic gaps in the fossil record.[52]

Next, Betz was given the opportunity to counter Gish's presentation. Betz made it clear, just as he had during the 1974 debate, that he was both an evolutionist and a creationist. He felt that it was quite possible that God created all of the natural laws that are in operation today, including evolution. The conflict between science and religion Gish alluded to was not necessary and did not reflect all religious people's views, he cautioned, pointing to himself as a prime example. Betz said that in some ways there was common ground between himself and Gish in that evolution certainly should be scrutinized and not held up as the one and only answer. During his speech, Betz was conciliatory toward Gish, not wanting to be painted as a mere enemy. But he made it clear that Gish's stance would introduce religion into the science classroom, where it didn't belong.

Other people in the audience wanted to speak about the issue, but the chairman declared that it would be better to wait until another time. Other school board members generally agreed and unanimously decided to discuss the matter fully at a meeting the following month.

The April 22 meeting was a marathon. A flood of people were in attendance to make their opinions heard. In the preceding weeks, the board had been buried under a deluge of letters and petitions. According to the minutes, some of the more interesting items they received included the following:

- Statement from the Tampa Jewish Federation standing firm in its endorsement of the principle of separation of church and state and opposition to teaching of material in the realm of religious doctrine as a part of the science curriculum in the public schools system.
- Petition with twenty signatures of members of the Loyalty Class of Idlewild Baptist Church of Tampa requesting the teaching of creation in schools as given in the Bible.
- Eleven petitions from science teachers in 11 senior high schools with a total of 79 signatures, out of a possible 106 science teachers in senior high schools of Hillsborough County expressing support of the resolution of the National Science Teachers' Association and the Florida Association of Science Supervisors (which opposed mandating the teaching of non-valid science theories) and resolving that creationism is a non-science theory.[53]

Once the public-comment period was opened, thirty-two people took turns addressing the school board. There was a mix of scientists, pastors, students, parents, teachers, and concerned citizens. The comments ran the gamut of appeals to emotion, appeals to reason, and appeals to public opinion. There was plenty of finger pointing as various speakers sought to correct the statements of previous speakers. When the comment period was finally closed, eleven people had spoken on behalf of evolution, eighteen had spoken on behalf of creationism, and a few had offered cryptic statements that were hard to place in either camp.[54]

Finally, Superintendent Shelton was given an opportunity to report to the board the results of his staff's research into the matter. He explained that there was a fundamental disagreement between supporters of evolution and supporters of creationism: evolution relies on natural laws, whereas creationism explains origin in terms of a supernatural creator. "Evolutionists object to the teaching of creationism as a part of science curriculum because the theory was not derived from the scientific method," Shelton said.[55]

The board had previously instructed Shelton to figure out the best way to incorporate creationism into Hillsborough's schools. To do so, he based his research on a broad question: "What are the appropriate disciplines for the inclusion of both theories in the curriculum?"[56] It was clear that he and his school science supervisors were dead set against teaching creationism in science classes. But that conviction was put to the test against a determined school board that disagreed. Shelton deployed various tactics to please his bosses while remaining true to his conviction that creationism was a religious view that would plow right through the wall of separation between church and state. He made it clear to the board that he was working to find a place for creationism in the overall school curriculum—more or less in keeping with their wishes—but that he was trying to keep it out of the science classroom.

As directed by the school board, Shelton offered a couple of position statements for the members to consider:

> In the area of biology, the Hillsborough County Instructional Program will continue to emphasize those theories derived by the scientific process and recognized by the scientific community as appropriate for science. Evolution will be taught as a theory, not as a fact, and the nature of a theory will be thoroughly taught.

Teaching of non-scientific theories of origins such as creationism will be accomplished in courses designed to explore man and his ideas.[57]

Based on these position statements, Shelton recommended that a committee of science teachers, social studies teachers, and lay citizens be formed and tasked with determining where in the overall curriculum "non-scientific theories" would be taught. He added that any teacher who would be required to teach creationism would need special training concerning the teaching of controversial materials.[58]

The purpose of the April 22 meeting was to allow the public to have its say. The board members refrained from making any comments, and they agreed to study the information presented and make a decision in a week. The meeting was finally adjourned at five minutes till midnight.

"Cause the Nation to Fall"

One week later the board was back in session. Before the evolution subject was addressed, though, the school board handled other science-related chores. Ironically, the governor had proclaimed May 2–3 as Hillsborough County Science Connection Days. The meeting minutes stated: "The Board concurred with the governor's Proclamation on Hillsborough County Science Connection Days in consideration of the work of the Science Steering Committee of Hillsborough County and its efforts to increase the appreciation of science and science education."[59] The board was then treated to a presentation from Marsh, the secondary science supervisor, concerning the Science Olympics, and a group of elementary students performed a science demonstration. Chairman Hill thanked the students and then moved on to the main event. Shelton again presented the position statements and recommendations from the previous week's meeting and then braced himself for a tongue lashing. He didn't have to wait long.

Lewis was the first to wag his finger at Shelton and his staff. He took offense at creationism being referred to as non-scientific. It was inconceivable to him that the staff had done so much research and yet come to such an erroneous conclusion. Evolution was the idea that was unscientific and untrue, he said. He didn't want creationism relegated to an elective class that no one had to take. Something that stuck out to Lewis during the previous week's public comments was how some speakers claimed that inserting creationism into science courses would violate separation of church and state but that allowing creationism to be taught in other courses would be acceptable. To Lewis, this demonstrated that those speakers were not being objective and had ulterior motives. He scolded Shelton for giving the board a position statement that was simply a reworded version of one the board had soundly rejected before.[60]

Other board members then piled on. Newsome said that creation or evolution could be a science or a religion, and it was biased to say that one is scientific and the other is not. Rodgers thought that this was by far the most important decision ever to come before the school board. The minutes report: "Member Rodgers said she believed in this nation under God and felt strongly that continued debate would separate the creator from the nation and would cause the nation to fall."[61] Rampello said he had read several creationist textbooks and couldn't understand how they could in any way violate the separation of church and state.

Shelton once again explained that evolution was based on natural laws, whereas creationism relied on reference to a divine creation, "which you cannot explain without reference to a designer, a planner, a creator, or a supernatural being." He then made an interesting disclosure: "Superintendent Shelton emphasized that he was a creationist, but as superintendent of schools he had to recommend a curriculum designed within the laws as interpreted by the courts in this country." Lewis disputed Shelton's view. He relentlessly pounded on evolution as a highly unscientific idea that was propped up by a handful of elite scientists. Essentially, any Ph.D. candidate who didn't accept evolution wouldn't attain a doctorate from those who controlled his or her fate, he said. As a result, evolution was firmly embedded in the halls of science without being challenged as the clear sham it was.[62]

Shelton and his staff had taken heavy criticism ever since the issue of creationism had come up, but he finally received support during this meeting. Lowry mirrored the superintendent's thinking in that he believed wholeheartedly in creation but didn't think it should be taught in a science course. Cecile W. Essrig also spoke up, saying that Shelton's proposed position statements were a positive, open-ended approach that could work. Hill also voiced his support for the position statements, saying educators should be trusted to do the jobs they're trained for.[63]

Marsh and Lewis then engaged in a heated debate. Marsh argued that creationism was not science and didn't belong in the science classroom, and Lewis claimed that there was no scientific evidence for evolution. The debate was then cut off by a motion to vote on acceptance of the superintendent's position statements. It failed 3–4.[64]

Lewis then presented a new motion:

> That the staff accept appropriate materials from interested publishers, scientists, and lay persons and analyze its instructional resource value in instituting a dual or multi-model approach to teaching about origins in science and other disciplines of the school system's elementary and secondary curriculum.
>
> That the staff development resources be applied to securing material and consultants to implement this study and that the board receive reports on the staff's progress until a satisfactory program is available to elementary and secondary students in the classroom, not later than September, 1981.[65]

Staff members in charge of elementary education curriculum quickly pointed out that evolution was not discussed at all at their level. Lewis agreed to strike reference to elementary schools but warned that it would be put back if any evidence ever surfaced that evolution was actually taught there.[66]

A vote was then held on this new motion. Lewis, Rodgers, and Newsome voted yes. Essrig, Hill, and Lowry voted no. With a tie vote on Rampello's shoulders, the audience in the packed room was anxiously silent as he considered his decision. Finally, he said yes. "Shouts of Amen shattered the silence. Then thunderous applause filled the room."[67]

With that vote, Hillsborough County jumped into the national spotlight. *Time* magazine later reported:

> Over opposition from 90% of the local senior high school science teachers, the seven-member Hillsborough County school board in Tampa, Fla., decided to require science classroom time for theories that challenge evolution. Says John Betz, associate professor of biology at the University of South Florida: "These people think

evolution is essentially an immoral idea that gives rise to immoral conduct." Tampa's teachers "are incredulous," says Betz. "They can't believe it is happening."[68]

Betz clearly wasn't happy with the decision. "I'm certainly not against the idea of creationism," he said. "I'm certainly not against the teaching of it in schools. But the way scientific creationism has been defined, it is a perversion of science and perversion of creationism. It's an obscenity."[69]

"Are You Satisfied to Brainwash Children?"

Representative Bush was excited that Hillsborough County had mandated the teaching of scientific creationism, as he felt it gave his second attempt at a scientific creationism bill the ammunition it needed to pass. "It's very helpful to me, in that it establishes precedent," he said. "I will now be able to go to the Legislature and say this theory's being taught in schools in Florida." In April 1980, Bush introduced House Bill 107, titled "Balanced Treatment for Scientific Creationism and Evolution Act," which was essentially the same bill he had filed the previous year.[70] Speakers who testified on behalf of the bill during committee meetings explained that natural geological features seen today could be explained by a catastrophic flood and that the earth is only thousands rather than billions of years old. It's unknown whether the lawmakers were swayed by these arguments or merely felt the heat from mountains of mail and persistent creationist textbook lobbyists, but a House education committee approved the bill 7–6. "If we are truly interested in truth, and do not know what truth is, we should teach all credible theories," Bush said.[71]

According to some lawmakers, the bill not only crossed the church-state divide but also interfered with local control of schools. Rep. Richard Hodes believed that it wasn't the legislature's place to force specific curriculum on the school districts. Hodes pointed to Hillsborough County's decision as support for his view rather than support for the bill as Bush hoped. "I think it's a local decision, which Hillsborough County reached, and that's fine," he said.[72] Unfortunately for Bush and his supporters, there were several committees to navigate and his bill fizzled out. He didn't give up, though. He tacked an amendment onto an instructional materials bill:

> HB 776, Amendment 4—on page 2, line 8, strike all and insert New Subsection 3: The Council is herein directed to review, evaluate and pursuant to the provisions of Chapter 233.09 make recommendations to the Department of Education for the purchase of instructional materials that give instruction in Scientific Creationism.[73]

Despite some support for the amendment, it didn't make it into law.

In June, Winslow tried to ride the wave started in Hillsborough County into Manatee County and found cautious enthusiasm. The Manatee County School Board asked their lawyer's advice and was told that if the Bible was used as an educational resource in any way, the Supreme Court would see that as religion and thus unconstitutional.[74] Nonetheless, the board was hopeful. They decided to hold off on making a decision until they saw how things went in Hillsborough County.

But Winslow persisted. In September, he asked the Manatee County School Board once again to consider teaching creationism, and they finally agreed. An official procedure was crafted and presented to the board:

> Procedure 400-012 Biology Curriculum.
> Recognizing that there are two major theories relating to the origins of life, the Manatee County school system will include in its biology curriculum scientific evidence relating to the origins of living organisms and the interpretations of that evidence as presented in the theories of creation and evolution.[75]

On a vote of 4–1 the procedure was approved, and Manatee County became the second Florida school district to allow the teaching of creationism in the classroom.[76] Unlike Hillsborough County, though, there didn't seem to be nearly as much fanfare. There apparently were no lengthy public hearings.

The lone dissenting voice on the Manatee County School Board, Marge Kinnan, demanded during an October meeting that any scientific creationism curriculum developed would need to come before the board for approval and be open to public comment. She was concerned, because she had read in a newspaper article that the county's science education supervisor, Robert Kitzmiller, was a scientific creationist. Kinnan wanted to have a hand in selecting the people who would write the new curriculum, but she learned that Kitzmiller and five other teachers who wanted to teach creationism had already formed a committee for this purpose.[77] As soon as creationism proponents had been given the green light, they wasted no time, which caught Kinnan by surprise.

That September, with Manatee County marked off as a victory, Winslow moved on to Pinellas County. He had been hounding that school board about creationism for years. He asked them: "Are you satisfied to brainwash children?" Several other people also spoke, including a representative from the local branch of the Moral Majority, a newly formed national conservative Christian political organization. An article in the *St. Petersburg Evening Independent* reported on one parent's plea: "Winslow was backed by several ministers and parents who want the theory of creation taught in schools. Bruce Love, a parent, said he believes not teaching creation to his children is 'a violation of my constitutional rights.' He added that teaching only the theory of evolution 'makes the father of my children look stupid' and begged the board 'not to degrade the family any further.'" Opposing the teaching of creationism in public schools was League of Women Voters representative Marilyn Baly. She cautioned the board about the legal problems that could arise regarding separation of church and state.[78]

When the meeting was turned back over to the board, a motion was made to direct the superintendent to research and report back on the feasibility of creating some alternative curricula. The motion failed 2–5. But that wasn't the end of it. The board's chairman, Rev. Arthur Libby Albers, used the muscle of his position to override the vote and authorize a study of the issue.[79]

No one else was quite as gung-ho as Albers, though. A few months later, Superintendent Sakkis admitted to the media that he had yet to conduct the study and had no plans to do so anytime soon. Opposition to incorporating creationism into lesson plans wasn't strictly based on religious concerns, either. School board member Betty Hamilton had a practical concern: "We would have to develop our own curriculum, which is very expensive. I don't know if we can spend taxpayer's money to develop a curriculum that is basically religious."[80]

"A Spirit of Compromise and Conciliation"

Hillsborough County's decision in the spring of 1980 to mandate creationism instruction in its schools encouraged citizens in other counties to go before their own school boards and ask to follow Hillsborough's lead. For instance, Pasco County had avoided the notice of the creationists despite the targeting of many of their neighbors. But with scientific creationism in the works across their southern border, the creationists came knocking.

It started in February 1981 when Rev. Lewis Turner, a local pastor who also headed the area's Moral Majority chapter, decided that evolution had gone unchallenged for too long in public schools. The *St. Petersburg Times* reported:

> "There are those who say that Darwin's theory is not a theory at all, but an established fact—as well established as the fact that the Earth is round," Turner said. "But there's as much credible evidence on the other side to show that life came about through spontaneous creation." He declined to elaborate.[1]

Despite the chorus of Turner's supporters imploring the school board to take action, the members opted for the politician's refuge: they took the matter under consideration. Turner found that response promising and felt that the board had "an air of cooperation."[2]

Nonetheless, plenty of skepticism was reported. Superintendent Thomas Weightman expressed his worry that complicated legal action would ensue should Pasco County mandate the teaching of creationism. School board attorney Joe A. McClain predicted that teaching creationism would open the door to countless other origin theories. "And I think we would be hard pressed to find the student who could carry the textbook," he said.[3]

When creationism flares at the school board level, it is often the case that a teacher or two comes forward to say that creationism is already being taught in some classrooms. A retired Dade County biology teacher told the Pasco County School Board members that throughout her career she had taught creationism alongside evolution. She also knew many of her fellow teachers did the same.[4] At that time Dade County had no official policy on the matter.

In June, another pastor, Neil Sanford, took a crack at it, but he was rebuffed by the Pasco County School Board. This time the textbook angle was attempted, with Sanford asking for equal space for creationism in their pages. There was a brief discussion about censorship concerns. The board's attorney assured the members that barring creationist texts was not censorship.[5]

The issue cropped up again in November when Sanford returned with *Biology: A Search for Order in Complexity* in hand. He asked that the board adopt it for use in schools. The request was passed along to a textbook review committee.[6]

"I Am Not Anti-Religion"

Meanwhile, the Manatee County School Board was having second thoughts about the new biology curriculum procedure they had adopted in September 1980. A progress report brought to the board in March 1981 showed that a committee of biology teachers was diligently working on a curriculum but that it wouldn't be ready for implementation during the current school year. The board voted unanimously to withdraw the new official procedure for the time being and reconsider it at a future date when the curriculum shaped up. According to the meeting minutes, Superintendent William Bashaw stated that "most people in the community feel the Board went into this subject rather prematurely without sufficient study but that there will have to eventually be a solution to the problem."[7]

The board tackled that problem again in November 1981. Finally, the committee of biology teachers proposed a policy for handling the issue of "origins" in Manatee County schools. The teachers suggested that the whole issue be dramatically simplified. When it was time to address "origins" in the classroom, a short statement about evolution would be balanced by a short statement about creationism. The policy then says:

> In recognition of the fact that origins of life take up less than one percent of the biology curriculum, and because of the developing and unsettled national legal situation regarding this subject, teachers should utilize the explanations herein described. It is suggested that when students seek expanded viewpoints or explanations of these statements, they be referred to their parents or religious counselor.[8]

The man responsible for getting everyone to talk about creationism in the first place, retired Rev. Clarence E. Winslow, was on hand to prod things along. He was given yet another chance to state his case to the board and field their questions, but he didn't have as much influence this time.

The board's waning confidence in Winslow showed in the skeptical questions and statements that followed. Board member S. Lyman Chennault asked Robert Kitzmiller, the supervisor of science education, if evolution was currently in the curriculum. Kitzmiller said there was a chapter in the textbook that focused on evolution and that the subject was addressed in other chapters as well. However, he admitted that because the issue was so controversial, many teachers skipped it. Board member Betty Nevin didn't like the proposed policy at all. The meeting minutes quoted her: "I have no objection in having the teacher state that there are other theories and among them are scientific creationism and they could read a paragraph and advise the students that we have material available if they want to pursue that but I never intended for that to replace the study of evolution." The majority of board members agreed with Nevin. She made a motion to kill the proposed policy, resulting in a 3–1 vote for rejection.[9] Manatee County was now out of the creationism spotlight.

Winslow tried his luck again in Pinellas County in February 1981 after getting nowhere there a year ago. Pinellas had rebuffed his attempts to mandate the teaching of creationism, but it had permitted teachers to present the subject in class since at least the early 1970s. Back then Winslow had presented to the school district creationist textbooks, which teachers were welcome to use. That wasn't good enough for Winslow, as everyone in the area knew by now.

Superintendent Gus Sakkis wasn't enthusiastic about creationism, having put off the school board chairman's direction to research the subject in 1980. Now that Winslow had brought the issue up yet again, Sakkis was even more opposed to it. In a memorandum to the board he recommended that the current policy of allowing but not mandating creationism be continued. He wrote:

> I firmly believe in the doctrine of separation of church and state. The request for mandating the teaching of scientific creationism is not a simple two-model, equal time request. The basis of scientific creationism is the fundamentalist Christian religion.
>
> At this point, I wish to make it clear that I am not anti-religion and the issue is not one of religion vs. anti-religion. The issue is should one specific religion be mandated to be presented to the exclusion of other religions and to children of other faiths. I believe not.[10]

The board eventually agreed with Sakkis, handing Winslow another defeat.

"Somebody Has To Teach Morals"

There were several efforts in the early 1980s to incorporate scientific creationism into public schools in spots all around the state. In March 1981, a panhandle county joined the fight. Okaloosa County superintendent Max Bruner Jr. was eager to get scientific creationism into his schools, but he had to contend with a 1975 federal court order banning the school system from having daily prayers and Bible readings. Bruner had the support of the head of the Okaloosa Moral Majority, Rev. Ken

Martin, who said, "It's not a matter of forcing anything on anyone. We believe both sides should be able to practice what they believe." A few months later, though, Bruner quietly gave up, saying he would leave it to the state legislature to get scientific creationism into the schools.[11]

There were rumblings in Polk County on the subject, but they remained in the background. The local chapter of the Moral Majority considered pushing for scientific creationism, but there is no record they ever went through with it. School district staff hoped to avoid the controversy, preferring to wait and see how things fared in Hillsborough County while also expressing concerns of their own: the cost of changing the curriculum, finding time to teach it in an already full school year, teacher training, and the possible appearance of teaching religion.[12]

Further north, a Marion County group called Citizens for Morality was organized in early 1981 with the purpose of being a "vehicle for moral citizens to unite and let their beliefs be known." One of the group's biggest targets was the public school system, where they wanted voluntary prayer, Bible readings, and "creationism and traditional family values" in the classroom.[13] However, their campaign met with resistance from the school board. When they asked for scientific creationism to be added to the curriculum, the school district gave them a clear answer: no. A statement issued by the district explained they were reluctant "to teach the Bible due to past court cases, and the absence of any guidelines as to an agreement among creationists as to exactly the way it should be taught."[14]

Despite the roadblock, Citizens for Morality persisted. "There is no scientific proof that man evolved from animals; none," said the group's president, Lewis Dinkins. "The theory that God created man is more plausible and more acceptable." The group believed that scientific creationism was an important link to further character education. "Somebody has to teach morals," Dinkins said.[15] However, they never accomplished their goals in the Marion County schools.

More than one hundred Volusia County science teachers, school administrators, and principals watched a presentation in Daytona Beach in March 1980 by the California-based Institute for Creation Research (ICR). The group's director of curriculum development, Richard Bliss, told them about his organization's two-model approach to teaching evolution. It all made sense to Superintendent Donald Gill and Assistant

Superintendent for Instruction Robert McDermott. There was a lot of good evidence supporting the creation side, McDermott said, while Gill commented that the two-model approach had merit. Both men cautioned that the issue would have to be studied and that no decision could be made for quite a while. Although Gill and McDermott were fans of creationism, some of the county's teachers had the opposite reaction. The chair of Mainland High School's science department, Kathleen Anderson, said, "I thought they were rather offensive. I feel they misrepresented Darwin's theory of evolution and misrepresented what the teachers are doing."[16]

The ICR also took their presentation to legislators in Tallahassee. However, the *Daytona Beach Morning Journal* dogged them there when the newspaper discovered that some of Bliss's materials "viciously" misrepresented the views of famous paleontologist Stephen Jay Gould and David Raup, chairman of the geology department of the Chicago Field Museum of Natural History. Gould and Raup told the reporter how the ICR distorted and lied about their views, to which Bliss responded, "But I frankly don't know why Gould gets so excited."[17]

The Volusia Association of Science Teachers was wary of the ICR's visit to the state legislature. The organization preemptively adopted a position statement against possible anti-evolution legislation. The statement said the association "is opposed to the outside interjection of any controversial science curriculum into Volusia County unless such proposed curriculum has been accepted locally after a carefully planned workshop for study, open discussion, uncensored questioning, examination of materials, and full presentation of all viewpoints."[18]

Amid all of the creationism commotion in 1981, Senators Dan Jenkins from Jacksonville and Alan Trask from Winter Haven introduced a creationism bill in their chamber while Rep. Tom Bush made his third annual attempt in the House. Both bills foundered in committee, but that didn't stop the lawmakers. They attempted to tack creationism onto other bills, such as a bill concerning the disposal of old textbooks. As his amendment was debated, Jenkins said, "There's two major theories regarding creation today and only one of them is being taught. I think it's a sad commentary when 70 percent of Americans believe the other way."[19] It was a futile gesture, as the amendment was ruled out of order and the teaching of creationism overall died yet again on the state level.

"Touted Loudly throughout the Land"

Even though most other counties weren't diving into the roiling creationism sea, Hillsborough County was in over its head. Nancy Marsh, the county's secondary science supervisor, had previously taught high school biology and had a master's degree in microbiology from the University of Florida, but nothing prepared her for the shark-infested waters she found herself in during the early 1980s.

She says that she had never heard of the ICR before then, and she and other staff members initially thought the creationism push was an honest local effort led by Hillsborough citizens.[20] It wasn't until later that she found out it was actually a national movement led by the ICR, which saw her willing school board and large school district as just another stepping stone toward wider recognition for its cause.

Marsh had plenty of time to get acquainted with the ins and outs of creationism as she spent more than a year and a half deeply immersed in spirited debates while helping hammer out a brand-new origins curriculum. The task was such a monster that after several months of origins-focused work Marsh wrote a memo to her superintendent outlining the many other science supervisor tasks she had been forced to neglect over time.[21]

Marsh and all the other involved supervisors were buried under mountains of correspondence. For instance, right after the school board officially required the teaching of creationism at their April 1980 meeting, Marsh received a letter from Dr. Wayne Moyer, executive director of the National Association of Biology Teachers. Moyer wrote: "I am dismayed to learn of Hillsboro [sic] County's school board's decision to include creationism in its curriculum. It will be touted loudly throughout the land as a victory for creationists, and used to pressure wavering school boards. The danger, of course, is that creationism is being made equal to scientific study of evolution, even though both are based on radically different assumptions, and cannot possibly be compared to each other."[22]

People came out of the woodwork offering opinions and assistance. Jehovah's Witnesses canvassed the county's schools, offering a special edition of their *Awake* magazine titled "Accidents of Evolution or Acts of Creation?" to principals who in turn forwarded them to Marsh. The publication was heavy-handed in its portrayal of the controversy, saying, "Many Christian parents feel their children are under attack. The target, their children's faith. The place of attack, the classroom. The attackers, evolutionists."[23]

Marsh was also in regular contact with other county school systems who wanted to know what was going on in Hillsborough County, and organizations like the PTA wanted to be kept up to date. Textbook publishers asked for Marsh's advice, such as a Ginn and Company science publications editor who offered to send chapter manuscripts about the "origin of life and on evolution" to Marsh. "Your comments on classroom acceptability as well as the pedagogic desirability of approach would be most helpful."[24]

Marsh needed a friend to lean on during these stressful times, and she found one in a former professor of hers, Dr. John Betz, a microbiologist at the University of South Florida. Betz was already familiar with the national creationist movement, having debated its leading lights in 1974 and then again during a Hillsborough County School Board meeting in 1980.

Betz's previous arguments had fallen on deaf ears. Once the teaching of creationism was mandated in the local public schools, he became a curriculum adviser and committee member fighting hard to keep the scientific creationism lessons from crossing any lines. Betz labored to educate the general public about evolution. He was frequently quoted in local newspapers and even the national media, and he wrote a feature-length article for *Tampa Magazine* about creationist tactics. His article unapologetically pointed an accusatory finger at creationists for causing strife in the classroom: "But why worry about court suits over a simple question of what should be taught in a science class? Didn't that all end years ago in the famous Scopes Monkey Trial in Tennessee? In this enlightened scientific age, when Homo sapiens has found a cure for polio and put a man on the moon, is it possible that children's science classrooms will be the battlefields between squabbling proponents of this idea and that? The answer, sadly, is yes."[25]

"Irrational Impossibility"

Now the monumental task of creating the actual curriculum had to be completed. The time frame was relatively short, allowing only about sixteen months from initial planning to pilot program implementation. Various committees were assembled with an eye toward balancing the participants' views as much as possible. Professors from the University of South Florida biology and religious studies departments were invited to be committee members and advisers. Faculty from Temple Terrace's Florida College, a Bible college, provided a counterbalance. Eventually, the Hillsborough County School Board appointed a forty-two-member Origins Curriculum Committee that held its first meeting in late October 1980. The committee included teachers, local business representatives, a rabbi, citizens, and representatives of the ICR.

The committee's first order of business was to establish goals and guidelines that could be passed down to a curriculum-writing subcommittee. It took seven meetings to piece together six goals and twenty-one guidelines. It was a slow, uphill battle, as evidenced by the guidelines' introductory paragraph: "It is obvious that we disagree deeply and on many points. But in order to do something positive and constructive we must find common ground, discover and define things we can agree on, as a basis both for our further discussion and, ultimately, for the instruction of students regarding these issues. The following are offered in a spirit of compromise and conciliation as a possible basis for proceeding affirmatively."[28]

The twenty-one guidelines boiled down to a plea for respecting opposing viewpoints. But the issue was so contentious that the ground rules for conduct had to be specifically spelled out. For example, this sequence essentially repeats the same statement but with a different target audience for each.

> 15. Let us agree, and teach, that it is at least conceivable to some that creation could have occurred.
> 16. Let us agree, and teach, that it is at least conceivable to some that evolution could have occurred.
> 17. Let us agree, and teach, that it is at least conceivable to some that both creation and evolution could have occurred.[27]

However, even this attempt at building bridges was unstable. Committee member Rabbi Martin Sandberg felt that the goals' promotion of a dual approach to origins was an "irrational impossibility" and would inevitably invite the teaching of religious doctrine in the public schools.[28] He warned that a replay of the Scopes trial was on the horizon. Needless to say, there were many heated discussions anytime the goals and guidelines committee met.

The committee presented their finished document to the school board for approval on December 16, 1980, but in doing so a significant rift in the committee was exposed. Seven committee members presented a "Partial Dissent" letter to the board, explaining how they were struggling with what they perceived to be the creationist committee members' religious motivations.[29]

The dissent was sparked by guideline 13, which originally stated: "In an academic context, creation should be taught as a man-made concept to explain human observations to the human mind."[30] The creationist faction of the curriculum committee found that phrasing unacceptable. In their view, creation is divinely revealed, not man-made. However, everyone eventually agreed to replace the phrase "man-made" with the word "scientific."

Despite the compromise, the conflict alarmed other members of the committee, including Betz. In the dissent letter describing how the incident unfolded, they explained to the school board the context of that change. The committee's creationists had been steadfast in denying that scientific creationism could have been conceived without divine revelation. Betz and other committee members countered that referring to scientific creationism as divinely inspired was blatantly unscientific and would give creationist concepts "the color of being more moral, ethical, righteous and Godly than evolution."[31]

Creationists had originally sold their idea of including scientific creationism in the curriculum to the school board by saying it could be done through a strictly scientific method. However, Betz saw firsthand evidence that religion was, in fact, a significant element of the creationists' position. The purpose of the dissent letter was to make that publicly known, since the final wording of the disputed guideline masked the issue.

Betz's views were even validated during his presentation to the board members. Arnold Schnabel, a member of the committee, spoke up to say that using the word "scientific" in that guideline was a good compromise. "It could come from God, it could come from man," he said.[32] With that admission uttered right there before the school board, Betz tried to make clear that he wasn't just nitpicking over a simple word choice. He requested that another statement be added to the guidelines document which explicitly stated that creationism was not to be taught as a revealed concept.

The board members seemed to be nonplussed by the committee's internal debate. They chalked up the arguments as standard fare with the opposing sides both eventually compromising and producing a product fair to all. With a collective shrug of the shoulders, the board thanked the committee for their work so far and approved the goals and guidelines with the word "scientific" in guideline 13. "Some people lost their argument and refuse to live with the compromise they should live with," board member Roland Lewis said.[33]

This was merely the first step toward a classroom curriculum, and there was already blood in the water.

Other difficulties were constantly popping up throughout the contentious process. Hillsborough County's general director of secondary education, Sam Horton, needed to ask the school district attorney for legal advice:

Since creationism implies a creator, would reference to a creator i.e. God, Muhammad, etc. violate first amendment rights or separation of church and state?

What happens if parents refuse to enroll students in a required course in which the creationist view of origins is mandated?

What happens to a teacher if he refuses to teach the creationist view of origins?[34]

Attorney W. Crosby Few responded that the First Amendment does not prohibit schools from referring to or teaching about various religions; instead, it prohibits espousal of any particular religion over another. If the class was within this guideline, then the parents couldn't refuse to sign kids up and teachers couldn't refuse to teach it.[35]

Even though the primary focus was on the science curriculum throughout this whole process, science wasn't the only academic subject affected. A committee was also formed to examine how the study of origins—as this subject matter was often referred to in Hillsborough County—would be addressed in social studies courses. It was eventually determined that although the subject was rarely addressed in social studies, teachers would still attend training sessions that "will center around increasing their knowledge" of origins materials. However, avoiding controversy in the classroom seemed to be the priority. "The future approach taken by social studies teachers would be as it is now being done with the teacher not taking a position on origins in order not to infringe on a student's beliefs and leaving open any and all possible explanations," a social studies supervisor advised in a memo.[36]

Multi-Model Approach

Now that a set of goals and guidelines was in place, the next step taken through the curriculum minefield was the actual writing, despite the undercurrents of lingering controversy. The curriculum writers added a whole new twist to the saga, coming up with a proposed lesson plan that instead of a dual-model approach—evolution alongside scientific creationism—adopted a multi-model approach. Scientific creationism would share classroom time with neo-Darwinism, punctuated equilibrium, and creative (or theistic) evolution.

When this approach went public, it predictably didn't sit well with some professionals. William Mayer, director of the Biological Sciences Curriculum Study in Boulder, Colorado, wrote a letter to science supervisor Marsh: "To see scientific creationism listed as a theory comes as a great shock. Whatever else scientific creationism is, it is not a theory. The four 'theories' listed really are not theories of origin. I have no idea what, for example, 'creative evolution' is supposed to mean. Punctuated equilibrium is one of the mechanisms of evolution. In short, the list of 'theories' comes across as completely incomprehensible."[37]

To gather as much information as they could for the curriculum, the writers took every opportunity to hear from prominent creationists. For instance, a field trip was organized to hear Dr. John N. Moore, a scientific creationist from Michigan State University, speak at Manatee

Junior College in Bradenton. Staff development funding from the school district was used to pay for an anticipated forty-five teachers to attend the evening event, with transportation provided by school buses.[38]

There wasn't a single element of the new curriculum that didn't result in debate. An early draft of the document referred to Darwin as a scientist, but creationists on the committee didn't like that. It was felt that since Darwin never obtained an advanced degree, he didn't deserve to be called a scientist. Instead, they insisted that he be called a naturalist and they won. Subsequent drafts replaced "scientist" with "naturalist."

Marsh and Director of Secondary Education Sam Horton had no problem with the compromise language, pointing out that it was a small issue compared with the other debates raging about the age of the Earth and the fossil record. But Betz hated to concede even on issues others saw as minor. He pointed out that using the creationists' reasoning, Albert Einstein and Isaac Newton wouldn't be scientists either. "Sure, Darwin had no advanced degree in science," Betz said. "But the guy's works spanned 30 years and he came up with some phenomenal findings. There's just no doubt 100 years later of his contributions to science."[39]

Be Open-Minded

Finally, a fifty-three-page curriculum document titled "The Study of Origins" was forged. It's a fascinating document that clearly reflects the tensions exhibited during its creation. Students are walked step-by-step through the tricky subject matter, starting with a pointed unit introduction that tries to establish ground rules about the nature of science. It cautions students that this subject is controversial, saying that some people believe that life stems from strictly natural processes, while others believe that life was sparked by the supernatural.

Students are told that this unit will stick strictly to scientific explanations that "use a controlled process of gathering data and drawing conclusions known as the scientific method." Then brief descriptions of the scientific method and scientific theory are provided. With that as a foundation, the lesson moves on to an explanation of the four concepts of origins. "Now, what kinds of explanations attempt to answer our original question of Origins? There are several very complex explanations including scientific, theological, and even philosophical explanations. However, since this is science, let's concentrate on the scientific explanations."[40]

The four concepts are briefly introduced, with more in-depth material to come later. The definitions are:

Neo-Darwinism: Life originated from non-life by physical-chemical processes under special conditions, and all organisms, both living and extinct, are related by descent with modification.

Punctuated Equilibrium: Life originated from non-life by physical-chemical processes under special conditions, and all organisms, both living and extinct, are related by descent with modification. The rate at which modification has occurred has been extremely rapid in some instances and nearly static in others. Most speciation results from rapid genetic modification.

Creative Evolution: A supernatural influence originated and may actively continue to maintain all of the matter and energy in the universe and the natural laws which govern it. This supernatural influence may continue to act in the universe. Some believe the supernatural influence created the natural processes of chemical and biological evolution and from these originated the diversity of living things on earth. Others believe that the diversity of living things may have been brought about partially by evolution and partially by subsequent creative acts.

Scientific Creationism: Life appeared on earth abruptly as various original and different groups of organisms in fully functional form. Genetic variability permits changes within populations (microevolution) but excludes changes from one major group to another (macroevolution).[41]

Next, students are instructed to identify and write down the assumptions, evidence, mechanisms, and limitations for each of the four concepts. They are told to use this information as a tool to help them explore the concepts in more detail. With a final caution to be open-minded, the students are then treated to a slide presentation that reinforces what they've learned so far.

Finally, they delve into detailed lessons on each of the four origins concepts with neo-Darwinism up first. There are eighteen pages devoted to evolution in the curriculum guide, including discussions of Darwin and his initial discoveries, examples of similarities in body structures seen in diverse living things that are evidence of common ancestry, and explanations of genetic variations and mutations. This is followed by a quick side trip to discuss a famous experiment showing how life could have arisen in a natural process from simple organic compounds. Then there is a primer on fossils and how scientists can tell how old they are.

Only five pages cover punctuated equilibrium. The lesson explains that species can remain unchanged for long periods of time and then suddenly go through rapid change under specific conditions. Interestingly, the text sets up punctuated equilibrium as separate and in opposition to neo-Darwinism when, in reality, it's a proposed explanation for how evolution happens.

Creative evolution gets the least coverage, taking up only three pages. "The evidence used to support Creative Evolution is the evidence for a finite beginning of the Universe and its matter, energy and natural laws."[42] The lesson goes on to explain that natural laws can be traced back to the Big Bang, but no further, which implies a possible supernatural influence. This section's summary heavily emphasizes that evolution and some form of creationism can coexist.

The lesson is wrapped up with a fifteen-page explanation of scientific creationism. The lesson states: "The extreme complexity of the living cell, the requirement of all systems to act in specific sequence, and the fact that there is no natural tendency to go from disorder to order or from simple to complex tends to support the idea of a master plan for scientific creationism." Students are told that some species of living things do experience changes but those changes are limited. For instance, genetic mutation experiments on the fruit fly do show the wide variations of characteristics possible in the species, but the fruit fly is still a fruit fly. Another section of this lesson focuses on the fossil record: "It is the contention of scientific creationists that catastrophism, in the form of one or more large-scaled floods, produced widespread sedimentation, fossilization, and extinction."[43]

When January 1982 rolled around, a solid draft of the curriculum was ready for review by the school board. A pilot program plan was prepared, including a timeline for implementation. In just one month, three high schools would give the new origins curriculum a test run.

But then it all came to a screeching halt.

Many Reasons for Delaying

Marsh recalls being called in to the superintendent's office and asked to read a federal court decision that had just been released out of Arkansas. In the case of *Rev. Bill McLean et al. v. The Arkansas Board of Education et al.*, it was determined that mandated balanced treatment for creation science was unconstitutional because it violated the First Amendment's prohibition of an establishment of religion. Marsh remembers that the court decision she read contained elements she immediately recognized from the work on Hillsborough County's new origins curriculum. Additionally, a similar court case was working its way through the federal court system in Louisiana, and scientific creationism was on the losing side there, too.

In the blink of an eye, the issue was once again before the Hillsborough County School Board. The local branch of the American Civil Liberties Union had already warned that it intended to take Hillsborough County to court if the new curriculum made it into the classroom, and the success in Arkansas now provided momentum for such a case.

Back in April 1980, school board member Sam Rampello's tie-breaking vote had officially launched the contentious curriculum-change process. Nearly two years later, Rampello, now the board's chairman, asked board attorney Few to study the unfolding story and report back to the school board at an emergency meeting his opinion on how the Arkansas decision would affect Hillsborough County's plans.

At the January 11 meeting Few had bad news for scientific creation proponents. He said that he had done a lot of reading and even consulted with assistant attorney generals in Arkansas and Louisiana. The audience heard from Few about U.S. District Judge William Overton's findings in the Arkansas case and how those decisions were directly related to Hillsborough County's plans.

As related in the meeting notes, Few reported: "Judge Overton found that scientific creationism was not merely similar to the literal interpretation of the Book of Genesis but rather that it was identical and parallel to no other story of creation. Further, Judge Overton held that creation of the world out of nothing is the ultimate religious statement because God is the only actor." Few compared the Arkansas materials to that produced by the Origins Curriculum Committee in Hillsborough. He found: "The definition of scientific creation as gleaned from the proposed curriculum on origins for Hillsborough County schools is substantially identical to that set forth in the Arkansas Statue."[44]

He went on to say that "about one-half of the material for possible inclusion in the local curriculum was published by the Institution for Creation Research which Judge Overton found to be affiliated with religious organizations. Nine of the eleven creationist writers acknowledged by Judge Overton, as being recognized as authorities in the field by other creationists, submitted materials which were used in the development of the local proposed origins curriculum."[45]

Few also noted "that the final origins curriculum guidelines for Hillsborough County provide that creationism should be taught in such a way that simply teaching about it should not contradict or support any religious beliefs. Judge Overton found that a similar requirement in the Arkansas Statute was self-contradictory in that teaching of creation implies the advancement of a religious belief."[46]

Finally, after relating to the school board and the audience his lengthy, thorough report, Few advised that Hillsborough County should put the origins curriculum pilot project on hold. The time and expense that an inevitable court case would suck up would be a heavy and unnecessary burden. Few said it would be best to wait and see how the Arkansas case fared in the appeals process, and to also keep an eye on the Louisiana case *Edwards v. Aguillard,* which was just getting started.

The responses from school board members were predictable. Rodgers felt that there were plenty of differences between the local plan and the Arkansas one. She wanted to move forward. Lewis insisted that the local effort was purely secular in purpose. He reflected on his time as chairman back in 1970. At that time the school board and school district staff had developed a thirty-page creation science guide for use in the classroom. It was ready to go before any religious-affiliated citizens ever got involved. The guide was used for some time and never caused a stir, which Lewis attributed to the guide's secular development.[47] He said the curriculum currently being discussed was in the same boat.

Citizens got a chance to speak after Few's presentation. Paul Antinori had been a state attorney of Hillsborough County in the mid-1960s. His opinion was that the Hillsborough situation would be found to be legal. The Arkansas case had featured many religious motivations out in the open for all to see. Antinori was confident that the local curriculum process didn't have the same underlying purpose of advancing religion weighing it down. He also pointed out that it is not illegal to use the Bible in the classroom for literary and historic purposes. The meeting notes say: "Mr. Antinori suggested, since the nation looks upon Florida and Hillsborough County as aggressive, the Board not adopt a 'wait and see' attitude but act unhesitatingly to give students a complete education."[48]

Next, Mimi Kehoe, representative of the League of Women Voters of Hillsborough County, avoided any discussion of religion, choosing instead to focus on dollars. She talked about how her group had worked closely with the school system for several years to secure funding for quality education. She felt that continuing to pursue the new curriculum would seriously harm all the good fiscal work that had been done.[49]

Bruce McKay told the school board that his personal historical research showed that Darwin was caught up in a grand fraud. McKay handed out copies of a rough draft of a book he was working on that was self-published several years later. The book was *Science Research Proves Evolution Hoax: The Conflagration, When Parallel Universes Merge.* The promotional materials touted the book as: "Packed with eye-opening facts that expose evolution as a faked ideology, 'the great apostasy' was set up using underground couriers and agents of pre-Nazi Germany, who plotted to reprogram the minds of the entire world!"[50]

Money became a big theme of the meeting with several people echoing Kehoe's concerns. Fred Rothenberg, president of an elementary school PTA in Tampa, was worried about finances and thought that parents should take their kids to church to learn about the "other theories." Sam Rosales, executive director of the Classroom Teachers Association, didn't have an official opinion on creationism but was concerned about the school board getting inappropriately involved in the writing of curriculum. He further explained that money spent on a lawsuit would be wasted, since the curriculum in question could be put on hold for years until the legal issues were finally resolved.[51] Board member Newsome confidently supported the curriculum, saying that there was no religious motivation behind it. He also felt that finances shouldn't dictate the school board's actions.

But then Lewis asked how much a lawsuit might cost. Few quoted anywhere from $50,000 to $500,000, depending on how long the process took and how far the school board desired to push it. Then the next obvious question arose: how likely was a lawsuit? Gail Davis, state board member of the Florida affiliate of the ACLU, was on hand to answer that one. She confirmed that her organization was ready to go to trial if the curriculum was finally approved for classroom use.[52] Backing her up was Dr. John Sellers, representing the American Chemical Society. He stated that the society had taken a clear stand against scientific creationism.

Lewis wasn't deterred. He took exception to several statements he found in the Arkansas ruling. He said that the line "Creationists view evolution as a source of society's ills" should actually be: "Some creationists view evolution as a source of society's ills."[53] Lewis was adamant that not all people with a religious background had religious motives when pushing for creationism in public schools. He insisted that his own motivation was to "get young people a balanced presentation of the volume of information which is available and being ignored." He felt that the Arkansas judge was illogical in his decision, mainly because he extrapolated from a few creationists the religious motivations of all creationists.

Board member Palomino countered that the same creationists the judge referred to as having primarily religious motivations had contributed significantly to the local curriculum. He felt that the different theories of origin could be addressed in an elective course rather than in a required science course. One problem he had was that scientific creationism started with a conclusion and then tried to find the data to support that conclusion, which is the complete opposite of how true science is conducted.[54]

Board members Lowry and Essrig said they would vote to defer teaching of origins. They personally supported creationism to some degree, but they didn't think it should be required in a public school. Essrig said, "It seemed there were many reasons for delaying the origins curriculum and not very many reasons to start immediately."[55] Finally, Lowry made an official motion to delay curriculum implementation.

Newsome tried to rally the other school board members by saying the

Hillsborough County School Board had never before taken the "wait and see" attitude and shouldn't start now. The curriculum development had been a two-year-long process that deserved to cross the finish line. Rodgers agreed with Newsome. She said that the local curriculum had been developed differently than in other places, and that when the school board originally voted to start the curriculum development process they knew full well that there might be a lawsuit. Lewis stated that the curriculum committee had worked hard to keep the project secular. He also pointed out that plenty of materials for the curriculum had come from sources other than creationists.[56]

A vote was called for, and those wanting to delay implementation of the origins curriculum won 4–3. Voting in favor of delay were Lowry, Essrig, Palomino, and Rampello. Voting against were Rodgers, Newsome, and Lewis.[57]

This vote meant that the new curriculum wouldn't move into the classroom as scheduled; instead, it could still be worked on in anticipation of possibly being used sometime down the road. Curriculum approval was on the agenda for the next meeting, but Newsome suggested that it be pulled now that they had voted to suspend its implementation. Lewis disagreed, saying that with continued work and review, the curriculum could be made "suit-proof." The school board then agreed to leave the origins curriculum on the agenda.[58]

That next meeting was uneventful. Marsh gave a presentation outlining all the components of the finalized origins curriculum. The school board thanked her and the committee for all the hard work put into the project.[59] And with that, scientific creationism struck its tents in Hillsborough County. The curriculum was filed away and never considered again.

4

"A History of Hoaxes, Deception, and Deceit"

With federal court cases in the early 1980s coming down hard on creationism in other states, it looked like evolution education also had won in Florida. The big fish, Hillsborough County, slipped away from the scientific creationists, and all the other counties that were nibbling at the bait scattered, too.

But Rev. Clarence E. Winslow was a patient, determined man. In June 1983, just a year and a half after Hillsborough County shelved its new creationism curriculum, Winslow was in front of the Manatee County School Board with his familiar sales pitch. Some board members were new to their jobs, and Winslow found a promising new convert to his cause among them. Now chairman of Citizens United for Responsible Education, Winslow made an impassioned plea to the school to give creationism room in the curriculum. He claimed that the data used to support evolution could just as readily be used to support creationism. He also fell back on an old argument he had been proclaiming for years: that the courts have never actually outlawed the teaching of creationism.[1]

New board member Louise Johnson seemed willing to get the ball rolling, saying she was "100% in favor of teaching both theories as the children should have the right to have both sides presented."[2] However, there was some confusion concerning how she could get it done. Because the school board had decided the matter a few years prior, it was believed that procedurally the only way to reconsider it was to have a board member from the winning side make a motion to do so. Since Johnson wasn't a board member back then, she was prevented from taking action.

That didn't sit well with some of the citizens who were in attendance. They demanded to know more about this procedural roadblock, prompting a lengthy conversation about what had happened the last time creationism was considered and what the board could do about it now. Audience members seemed to have the impression that the board was being dismissive of them. One man said that his children had been cheated out of learning about creationism in school, and he accused the board of dodging the issue. One emotional outburst was reported in the *Sarasota Herald-Tribune*: "Sharon Eddy, who said she is the mother of two children in the public school system, loudly told the board that 'Christians should have rights too.' She said the creationism issue is important to Christians, and then broke down in sobs. She told the board members that they are destroying children and that the world will end soon because of such actions." Chairwoman Marjorie Kinnan bristled at Eddy's remarks, saying that all of the board members considered themselves Christians.[3] Superintendent William Bashaw reminded those in attendance that the issue had been thoroughly discussed the last time it was before the board and that the public had had a big role then. Board members "do not operate in a vacuum," he said.[4]

It was clear that there had to be a way to officially discuss creationism again, but no one seemed to know how to go about it. One possible way might be if a board member wanted to "put a new item on the agenda which was not in fact the old agenda item, then that would only take a change of language or position and the board was certainly free to do that."[5] With the subject so unclear, the board voted to have the superintendent and school board attorney research it and return at a later board meeting with recommendations.

Before the next meeting, Johnson announced that she intended to bring the subject up. She felt that this was a matter of academic freedom. Either creationism should be brought in to balance evolution, or evolution needed to be banned. Surprisingly, Winslow said he would prefer that Johnson not bring up the subject. He had turned his attention to working behind the scenes with the new superintendent, Gene Witt, to find a way to include creationism. "There is more groundwork to be done," he said.[6]

Due to scheduling conflicts, it took a few months for creationism to officially appear on the board's docket again. In the meantime, the local branch of the ACLU sent a letter to the board advising against inserting creationism into the public school curriculum. This fired up one citizen, Jiri Taborsky, who felt the letter was inappropriate. He stood before the board during public-comment time at its August 30, 1983, meeting to challenge the ACLU's assertions. The creationism effort was not religiously motivated, he said, and the board should file an official complaint against the ACLU letter writer because the man wasn't actually a resident of the county. His lecture then took an interesting turn as recorded in the meeting minutes: "Mr. Taborsky then read from a pornographic magazine stating that the ACLU supported similar type magazines and that it defends, on the basis of the First Amendment, educational material which he would not like his children to see."[7]

When he was finally done, Taborsky gave the magazine to Kinnan. She later tried to give it back, but Taborsky told her to keep it. The superintendent and the board's attorney then assured the chairwoman that the magazine would be destroyed. Despite Taborsky's presentation, the creationism issue wasn't officially on that meeting's schedule and so no action was taken.

"Be Afraid of God"

But the subject was on the September 6 schedule, which attracted a predictably large crowd. More than 150 people were jammed into the meeting room, spilling out into the lobby. The school board's first task was to settle on an appropriate procedure for handling all the people who wanted to speak. First, thirty minutes would be divided among four people who had signed up ahead of time and were thus on the published agenda. Another thirty minutes would be afforded to audience members wishing to say something, with fifteen minutes being given to those on each side of the issue. Taborsky was in attendance, and he quickly complained that since several more people had shown up to speak for creationism than against it, the larger group should be given more time.[8] His opinion was duly noted but not acted upon.

Winslow was the first of the four scheduled speakers. He covered familiar ground, stating that the U.S. Supreme Court had never specifically ruled against the teaching of creationism, that humanism was becoming the state religion, and that there was rampant discrimination against those with traditional beliefs and values. These declarations were punctuated with frequent calls of "Amen" and "Hallelujah" from the audience.[9] Winslow pointed out that there were plenty of education materials about creationism now available and that students needed to be exposed to them in order to get a complete picture when studying the subject of origins. He requested that a committee be appointed to start work on a creationism curriculum.

Next up was Donald Brunner, who represented the local chapter of the ACLU. He briefly outlined for the board the legal history of attempts to challenge the teaching of evolution or to balance evolution with creationism. He assured the board that his organization would challenge any creationism instruction mandated in Manatee County schools. Rev. Myron Bunnell, who had been an ordained clergyman for sixty years and served at a local church for thirteen years, agreed that creationism should be kept out of the schools. It was a subject more fit for families' houses of worship, he said, because there were many different views about origins, depending on the families' personal beliefs. Bunnell felt that it was important for him speak up in order to remind everyone that many who oppose the teaching of creationism in schools are not atheists.[10]

Barbara Talburtt, president of the League of Women Voters of Manatee County, said that she didn't support one side or the other. Her purpose was to ensure some important questions were incorporated into the record and seriously considered:

(1) Are there legal ramifications to the addition of this study under the First Amendment of the United States Constitution, that is, the separation of church and state? (2) Are there funds available to cover the cost of possible lawsuits? (3) Will this program

be available on a voluntary basis or will it be mandated for all students? (4) What accommodations would be made for theories of origins from other religious beliefs?[11]

The four scheduled speakers had been able to lay out reasoned arguments for their views in the several minutes they were allotted. The speakers in the general audience, though, had little time to state their cases, and so resorted to more pointed and heated statements. One man warned, "Don't be afraid of the ACLU; be afraid of God," eliciting a rousing cheer from the audience.[12]

Angry that she had only fifty-three seconds to speak and that the board had instructed the audience to hold their applause, one woman invited anyone wanting to install a new school board to meet with her afterward. The audience was loud in their support of the emotional speakers, forcing Kinnan to call for order a few times. Another speaker challenged the board members with a Bible verse reading that essentially put the members on notice that "He who is not with Me is against Me."[13] Once the public comments wrapped up, eighteen people had spoken in favor of creationism and eleven against it.

Given the difficult task of delivering bad news in such an emotionally charged atmosphere was school board attorney Gavin O'Brien. The board asked him to present the results of his research into the matter. Having spent several days reviewing more than one hundred related legal cases, O'Brien was prepared with a thorough, hour-long presentation. He summarized a laundry list of court cases, and his conclusion was a sour pill for the creationists to swallow: public schools cannot teach Genesis as science in the science classroom. A snapshot of O'Brien's steep climb over the audience's mountain of resistance comes from the *Sarasota Herald-Tribune*:

> "In some cases, we are seeing the federal courts issuing permanent injunctions against school boards, forbidding them to readdress the issue (of creationism vs. evolution)," said School Board attorney Gavin O'Brien at Tuesday's board meeting.
> "That's Hitler!" shouted someone in the audience.
> "No ma'am," said the attorney. "Not Adolf Hitler—that's the third branch of the United States government."[14]

The law was clear: despite the personal preferences of the majority of the community, or even of board members, a public school system could not mandate the teaching of scientific creationism without facing the consequences of an expensive and lengthy lawsuit. O'Brien's command of the facts took the wind out of the creationists' sails.

After O'Brien was finished, Witt presented his recommendation to the school board. Witt was only three months into his new job as superintendent, but he had worked in the school district for nearly three decades and knew full well the issue's local history. A major problem he had faced with trying to incorporate creationism into the schools, Witt explained, was that there was no consensus on what the instruction would look like. In searching for a clear definition of creationism for classroom purposes, Witt realized that the matter was strictly a legal one. No single definition could satisfy even the majority of pro-creationists, he said, and the definitions that eventually wound up in the court system were stricken down.[15] "We may not have the luxury to do what we would personally desire," Witt said. "I do not find that I can recommend the board enter a legal battle, possibly spending thousands of taxpayer dollars, on an issue that has already been decided. I recommend you keep the status quo."[16] Without any speeches or dramatics, the board agreed with Witt's assessment. Creationists were again locked out of Manatee County schools.

A *Sarasota Herald-Tribune* article noted a conversation between Kinnan and Johnson that had taken place during a recess. Kinnan had expected Johnson to jump into the fray in support of creationism, and she was surprised when Johnson had remained silent. Johnson admitted that she had planned to make a motion to add creationism or remove evolution, but O'Brien's report convinced her not to.[17]

O'Brien's findings hadn't convinced everyone, though. The message Taborsky took from the creationists' defeat was that religion was barred from public education because those in power felt that it impeded the learning and thinking of students. To counter that, Taborsky founded the Student Science Research Program to help students create science projects that could incorporate religious beliefs. The stated purpose of the program was "to show that expression and knowledge of religious beliefs do not inhibit learning, but on the contrary, enhance the general knowledge, promote creative thinking, and contribute positively to the moral development of students."[18]

"Keep in Mind That These Are Theories"

As the issue of creationism in schools fizzled out on the county level, it was sparking to life on the state and national levels in the early 1980s. The tinder was textbooks, and it actually started with the intention to beef up textbook content, which many state governments complained had seriously deteriorated over the years. A 1984 *New York Times* article referenced a study that said "most textbooks present students with a highly simplified view of reality."[19] Textbooks in the elementary grades were considered excessively simplistic and even downright boring. The higher the grade level, the more likely that textbooks had been influenced by special-interest groups, especially in the subject of social studies. Of course, the controversy over evolution was another highlighted area.

Robert Graham, Florida's governor in 1984, consistently made education issues high priorities throughout his political career. When the winds of textbook change blew across the nation, he was on the forefront of the effort. He was quoted in the *New York Times* saying, "There's no point in providing the students with textbooks if they are little more than comic books." However, textbook publishers tended to tailor their nationally offered products to the needs of big states with large student populations, such as Texas. This led to the accusation that science education, specifically biology, was suffering across the country because in the mid-1970s those in charge of education in Texas had directed that evolution be downplayed as "only one of several explanations for the origin of humankind."[20]

That changed in 1984 when the Texas Board of Education reversed that evolution decree. People for the American Way coordinator Michael Hudson, whose group spearheaded the First Amendment campaign in Texas to make this change, told *Time* magazine: "This is going to free publishers to write about science accurately, unhampered by religious dogma. It undoes ten years of creationist influence on textbook content, and it will spill over into every state." And science education was just a symbolic start in the nationwide movement to overhaul textbooks. "States have upgraded requirements for graduation, raised teachers' salaries and enacted a variety of reforms," Graham said. "Parallel with these reforms must be a serious uplifting of the quality of textbooks."[21]

Florida's governor was so determined to force reform that he hosted a meeting in Tallahassee that year for influential publishers, legislators, and educators from twenty-two states. The goal was to figure out the best way for states to make textbook-buying decisions that would result in the highest-quality materials possible. The meeting of minds yielded mixed results. The participants were in agreement about the general purpose, but they balked at what some saw as Graham's heavy-handedness.[22]

In 1985 there was another backlash against attempts to undermine evolution's representation in textbooks. The California Board of Education voted unanimously to reject an entire list of textbooks because they believed that publishers had attempted to avoid controversy by glossing over or entirely omitting discussion of evolution.[23] While that was grabbing headlines, the same observation was made at the school district level in Florida. Textbook-selection committees across the state were in the process of rating textbooks that would be presented to the state board of education the next year for funding approval.

In October 1985, the *Miami Herald* published a story about a committee in Broward County whose members were surprised by how many textbooks glossed over evolution. The head of the science department at one of the county's high schools, Bill Tobias, was disappointed. "I think it's sad when book companies don't include evolution," he said. "I really don't care whether someone believes it or not, that's up to them. But I think they need to know something about it before they make a decision. They need background information for them to make up their minds."[24] The Broward County committee also noted that a book in use then in the middle schools, Laidlaw Brothers' second edition of Herbert A. Smith's *Exploring Living Things*, had a "to the student" notice printed inside that said, in part: "In this book we will explore some of the theories scientists have developed to explain the origins of life and the changes in living things. As you study, keep in mind that these are theories, not facts. Science will continue to work for final answers."[25]

Evolution was covered in *Exploring Living Things*, but brief arguments against evolution were also included. It was pointed out that "many people find it hard to believe that a cell could have come about by chance. Some believe that the earth and the living things on it were created as a part of a master plan." Later, students are told that there are other origin theories beside evolution: "People who believe in the creation of living things do not think new species can develop because of the changes that happen in organisms. They believe that each species is from an act of creation and that small differences will not change a species."[26]

"Institutional Child Abuse"

But as the events in Texas and California played out, and preparations for Florida's biology textbook selections coming up in 1986 got under way, a new movement was building in the Sunshine State. According to its leader, Shirley Correll, the movement's driving force was a fight against the promotion and establishment in public schools of humanism, which the teaching of evolution supported. Correll, of Polk County, led this charge on behalf of the Florida Action Committee for Education (FACE). She had first attracted attention in 1975 when she protested how textbook adoptions were conducted in Florida. She went before a state Senate Education Committee to report, "We are getting textbooks approved which promote homosexuality, perversion, the new morality and everything that we consider harmful to the home." The main targets at that time were social studies books and other writings that Correll said "promote hippies and radicals as heros [sic] and encourage leftist activism."[27] Included in her organization's list of complaints was an objection to teaching evolution as fact in the schools. However, that issue was further down on their priority list then.

Five years later, Correll was back in the news due to her group's protest against several textbooks that FACE claimed promoted "anti-Christian and anti-American beliefs." The *St. Petersburg Evening Independent* noted that Correll had been fighting for nearly a decade against what she considered the government's constant attacks on family values through the captive audience of students. "This assault is being carried out, she says, through the use of textbooks and courses filled with profanity, immoral sexual instruction, the theory of evolution and teachings against 'Family, God and Country.'" Correll appeared at a February 1980 meeting of the board of education, which was then composed of Governor Graham and members of his cabinet, to complain about books that were "hostile to the Bible" and "biased against religion." The cabinet agreed to postpone approval of nineteen contested textbooks until they had a chance to review the books for themselves. But when they met again in March, they had concluded that the books were fine and that Correll's arguments were "out of context and misleading."[28]

Correll was determined to continue her crusade despite these setbacks, devoting most of her time and energy to it. Some of her opponents labeled her as an extreme book censor who took text examples out of context and sensationalized them for effect. The grandmother saw herself as a defender of the family and morality. She went so far as to say that public schools' insistence on promoting humanism—which she defined as "inverted Christianity"—was "institutional child abuse."[29]

By 1981 Correll was traveling the state talking to local groups about her campaign and she became state director of the Pro-Family Forum. The next year she finally found success when she lobbied the state board of education to reject a textbook. The target was *Sociology* by University of Cincinnati professor Ronald Federico, because it claimed that homosexuality was an acceptable lifestyle, among other objections.[30]

With this victory, Correll's reputation grew along with her confidence. A *Palm Beach Post* feature article on Correll said that some teachers were intimidated by her and so took great care in watching what they taught in order to avoid any controversy that would attract her attention. Correll claimed that due to the commotion she made, no textbooks were reviewed the next year, because it was an election year and no one wanted to be in her crosshairs during such an important time. Whether that was true or not is debatable, but she definitely saw herself as a major player. "In the beginning, they thought I was a mosquito, a nuisance. They thought I would go away, but they have found that I am not going to go away, that my influence is expanding," she said.[31]

Correll eventually turned her attention to evolution, using it as an example of the humanism she was trying to root out of the public schools. Humanism was a recognized religion, she said, but the humanist teaching of evolution was getting special treatment over other religious ideas.

The *Ocala Star-Banner* reported on a 1983 meeting she spoke at in Marion County:

"All of a sudden, 'Bang,' there came life from the tooth fairy," she said, disputing the Big Bang theory of creation. "Life cannot come from nonlife," said Mrs. Correll, the wife of a chemist.

She decried Charles Darwin's theory. "All of us have probably had an ancestor swing . . . by the neck, not by his tail. I'll tell that evolutionist, 'get that monkey off our backs.'"[32]

Finally, as 1986 approached, the evolution tempest started brewing. Early in the year, biology textbooks would be considered for funding approval by Graham and his cabinet. The governor was solidly behind improving textbook quality, but he was also actively running for a seat in the U.S. Senate. Meanwhile, Florida school districts were pushing for more evolution in the textbooks, following in the footsteps of Texas and California.

Correll said that she was just waiting for the list of books to be handed over to the governor before she made any moves.[33] The February 19, 1986, board of education meeting did not disappoint. The textbook adoption meeting involved books for several subjects, but it was twenty-two biology texts and one home economics text that were the targets of presentations challenging them. The home economics text was troublesome because it covered premarital sex and cohabitation, while the biology textbooks were "dangerous" because they treated evolution as fact and neglected creationism.[34]

Concerned citizen James Brady stood before the governor and his cabinet with large poster boards displaying enlarged illustrations from some of the biology texts. Brady, who worked in an Escambia County schoolbook repository, pointed to the diagrams, claiming that they and other items found in biology books were examples of factual errors. Many of those errors were outright frauds that had been perpetuated for more than a century, he said. "A lot of material in these books is scientifically unreliable. The second thing is they're unpopular. They're against what the people want."[35] Next, a parent told the board that teaching evolution without the counterbalance of creationism confuses students and creates a rift between parents and their children. C. B.

Subrahmanyam, a biology professor at Florida A&M University, said that evolution should be presented as a theory rather than a fact; he went even further and suggested that children not be exposed to the theory of evolution until they got to college.[36]

The board then voted to postpone adoption of the disputed texts. Education commissioner Ralph Turlington assured everyone that such a vote was standard procedure anytime a controversy arose. Graham then instructed the panel that had recommended the books for approval to go back and take another look at the texts and see if there was any way to provide more balance. The chairwoman of the textbook panel, Donna Stull, was upset. "Creationism is a belief, not a science," she said, "and you put science in science textbooks. Evolution is a fact, the theory is how it happened, not that it happened."[37]

The anti-evolutionists were happy. They had already collected nineteen hundred signatures on a petition, and the publicity of the postponement stirred up yet more support for their cause. To top it all off, it looked like Graham was their ally when he said: "The purpose of school is to provide students a basis upon which they can make their own value choices and form their own opinions. I think schools ought to be places where all ideas are welcome and that students are led through a process (in which) they can determine themselves which is true." A few days later, though, the governor's thoughts on the subject were muddled. "I'm not proposing that we require the teaching of creationism," he said. On the other hand, he had no problem with presenting mature students with both ideas and letting them decide for themselves which had more merit.[38] That's all the anti-evolutionists wanted.

Stull, a science specialist for Broward County schools, faced a barrage of correspondence and suggestions meant to help her find a balance between evolution and creationism, as the governor had suggested. But she wound up dismissing it all, saying that the controversy "was old hat 20 years ago" and that she was surprised to be in the middle of this fight. She inspected the creationist-leaning materials offered and decided that "some of the books they recommended were weak in evolution and many other areas. They were not quality textbooks."[39]

"Protect Parents' Rights"

When the state board of education met again on March 5, the ranks of creationists before them had swelled. Among the loudest was Correll. "I feel as Moses must have felt when he went to pharaoh and asked, 'Let God's people go,'" she said. "Evolution implies there is no need for a creator. And the high rates of suicide show our children's hopelessness." She went on to call evolution the worst cover-up in history. A preacher told the board that the teaching of evolution was "a history of hoaxes, deception and deceit."[40] A parent presented petitions packed with more than eleven thousand signatures.

As a long line of speakers attacked the biology textbooks, the supportive crowd punctuated the speakers' remarks with amens and loud applause while sporting yellow buttons that demanded "Protect Parents' Rights." Two men were the only ones to speak up in favor of evolution, compared to at least a dozen speakers who passionately supported creationism. At least one of the pair was booed by audience members

Shirley Correll displays the publication *A Guide: Teaching Moral and Spiritual Values in Florida Schools* at a cabinet meeting in Tallahassee, March 5, 1986. (Mark Foley, State Archives of Florida.)

during his presentation. The meeting stretched on for two hours, but the creationist speakers didn't lose any steam. One parent said, "The format of these texts is always laid out from amoeba to man, no way else. We are fed up with the evolutionary establishment stuffing their religion down the throats of our children."[41]

When it was finally time for the board to make a decision, it was anticlimactic. With little comment, and no comment at all from Graham, the board voted unanimously to adopt all of the textbooks, including the home economics one. The crowd directed most of their anger at the governor, accusing him of having said things that creationists wanted to hear strictly for political gain while he was running for the U.S. Senate. Graham later defended himself, saying that the creationists had actually been fine with a handful of the adopted books and that local school boards had the freedom to choose their own textbooks, even those not approved by the state.[42]

Stull was generally satisfied with the end result, which didn't compromise her and the selection committee's education goals. "Creationism was not considered," she said. "It is a belief. It is not based on scientific evidence. If you're going to include beliefs, let's include witchcraft, UFOs, ESP. There are a lot of beliefs." However, Stull lamented that the textbooks available for her committee to choose from left a lot to be desired. Despite the new movement to strengthen textbooks, evolution had yet to get the coverage needed. "If the state committee had based its opinion on the texts' treatment of evolution, we wouldn't have put any books on the list," she said. "We ended up compromising. None of the books on our list have good treatments of evolution. One book doesn't even mention the word. They talk around it. All are using much more ambiguous terminology than texts used to use."[43]

Correll promised that her fight wasn't over, though. "We too have a dream, a dream of restored religious freedom in America. It cannot and will not be denied forever."[44]

"A Bunch of Bigots"

After losing their battle over textbooks at the state level, creationists simply stepped down to the school district level. Escambia County resident James Brady, who had spoken at the February 1986 state board of education meeting, witnessed a few months later his own local school board discuss several biology textbooks that he didn't like. The May school board meeting's agenda included adoption of textbooks for several subjects, but science dominated the conversation.

Several citizens offered impassioned opinions, leaving the board members with a weighty decision. Board member Ed Stanford felt the need to respond in kind. "For six months I have been listening to a lot of speeches and a lot of preaching," he said. "For the next three or four minutes I would like to do a little of it myself." He explained that he was a Christian who was very active in his church. Stanford had studied the issue for quite some time and then talked with proponents of both sides. "I tried to work out a compromise with each group so as to calm the troubled waters," he said, "but I regret to inform you that no one was willing to compromise one tiny bit." After more reflection, Stanford said, he had come to the realization that textbooks were actually a minor part of classroom instruction. Rather, the teachers are the key. "If you survey our teachers, I believe you will find 95% of them will say that they believe in God. Since the teacher is the most important factor in the teaching process, about 4 times as important as the textbook, the importance of one book over another has been grossly exaggerated."[45]

The board approved the disputed textbooks on a 3–1 vote.

Since there didn't seem to be any way to stop the disputed texts from getting into the classroom, Brady decided that the next best thing would be to offer a balance. He wrote a two-page handout titled "The Other Side of the Scientific Evidence Regarding Origins." Brady and members of his Accuracy in Textbooks Coalition presented the handout to the school board in July, asking that it be distributed to students in the schools. "No court or judge has ever ruled that you can't teach scientific evidence that casts doubt on evolution," he said.[46]

The handout challenged the teaching of evolution using seven points, such as questioning evidence for things like the age of the Earth, natural selection, and theories on the origin of life.[47] The overall purpose of the handout was to demonstrate that evolution is scientifically impossible.

Local biologist Lee McKnight attended the July meeting and wasn't impressed with the handout's arguments. He said: "The people who wrote this handout aren't scientifically trained. They are a bunch of bigots who are trying to undermine scientific theory because it goes against their beliefs." Speaking in support of the handout was Rev. Chuck Baldwin, pastor of Crossroad Baptist Church. Baldwin made it clear that the goal wasn't to eliminate the teaching of evolution but rather to provide balance, which, he said, is "what education is all about."[48]

Brady's coalition had a sympathetic ear on the school board: Chairman Jim Bailey. During the meeting, he made a motion to allow distribution of the handouts. "There's nothing wrong with giving two views," Bailey said. "It would be no more harmful than literature saying do not smoke cigarettes or do not smoke marijuana."[49]

School board attorney Lou Ray felt that distributing the handout in schools would open the district up to expensive legal challenges. He noted that the handout was thick with creationist language, which had recently been determined to be a religious view in federal courts. With that in mind, the school board declined to approve the handout on a split 2–2 vote. The board did offer to turn the handout over to a committee of parents and educators for them to determine if it would be appropriate for placement in school libraries. Brady was disappointed and felt that the offered compromise was inadequate.[50]

"Drop Our Bananas"

Just one month later, the issue of creationism was back in the news. Governor Graham was on his way to a U.S. Senate seat victory, and four Republican hopefuls were dueling in their primary race to replace him in the governor's mansion. The candidates—Bob Martinez, Lou Frey, Tom Gallagher, and Chester Clem—attended a forum in August where it was revealed that all of them supported the teaching of creationism in public schools. "I believe in the Catholic Church's teachings, and I have no reason to believe in an alternative," Martinez said. "Students should have the choice. What we're seeing now, all they're getting is one side." Gallagher agreed: "I believe in creationism, but I don't have a problem with other theories being taught also. There are many of us who believe the Bible is true and ought to be taught."[51]

Newspaper accounts emphasized that despite their heartfelt beliefs, the candidates were fuzzy on certain aspects of the controversy. Clem stated that there was scientific evidence supporting creationism, even though he declined to elaborate on what that evidence was. He was clear, though, that humans did not evolve from lower life-forms. Clem's running mate, Tom Bush, had served in the Florida House of Representatives in the late 1970s and early 1980s. He was well known then for his conservatism and for having offered bills promoting creationism several years in a row. While on the campaign trail with Clem it was clear his attitudes hadn't changed. "Chester Clem and I choose to believe we were created rather than one day we just decided to drop our bananas and climb down out of a tree," he said.[52]

When it was pointed out to Frey that a fundamentalist view of creation was in opposition to the theory of evolution, he insisted that the two views could be compatible while at the same time refusing to go into detail about his specific views of creationism. "You're getting into a whole philosophic, semantic area," he claimed. "I think you can believe in both. Come to church with me Sunday and we'll talk about it."[53]

As the gubernatorial race played out, creationism also popped up at the local level when a wave of Christian conservative candidates challenged what they saw as excessively liberal incumbents. The main stage for this drama was Palm Beach County, where a conservative group called the Committee for Responsible Citizenship actively attempted to recruit like-minded candidates by circulating a flier asking for qualified "Christian candidates." Once the campaigns got fully under way, though, the group regretted the very public way the recruitment played out, since it heavily colored how the race was perceived by the media and the public. It also prompted the formation of a group called Citizens for Quality Education, which fought back against what were seen as "fundamentalist threats."[54]

Moral and religious issues dominated the races, with school prayer, sex education, and creationism being the hot topics. Candidate Richard Lodwick said that he didn't like how religion had stolen the spotlight, but he also had a clear opinion about creationism: "A statement at the time (of the Scopes' 'monkey trial') was that if we didn't allow teaching of evolution in schools that was bigotry. Well, if it was bigotry then, I ask you, is it bigotry now? We're not allowed to teach creationism now. We can only teach evolution. It's still a closed system."[55] A few other school board hopefuls also supported creationism, but when the votes were finally counted on election day, all of the creationists were soundly defeated.

A veteran state senator seeking reelection, Don Childers, faced challenger Adele Messinger in a bruising campaign fight the likes of which Childers had never experienced before. Childers, from West Palm Beach, faced an attack on his personal creationist beliefs. The *Sun-Sentinel* reported on one exchange he had during a campaign stop when he was asked if he supported teaching creationism:

> Childers turned heads in the audience when he responded with a question of his own.
>
> "Does anyone here believe that they came from a fish?" he asked.
>
> Those at the political forum in the High Point condominium clubhouse in Delray Beach were momentarily silenced. Then, a couple of men in the audience taunted him, yelling "I did. I did." Childers didn't flinch.
>
> "I have a God. God created me," boomed Childers, the 54-year-old Skipperville, Ala., native who has lived in South Florida since 1960.[56]

However, Childers then separated his personal beliefs from his political job, saying that he never advocated teaching creationism in schools. He went on to win his race.

As in Palm Beach County, a group of conservative Christian candidates made the Broward County School Board elections exciting. Three political novices ran their campaigns on the promise of bringing "traditional" values to the schools. LaJune Lundquist, a grocery store bookkeeper, wasn't deterred by recent federal court cases which had ruled that creationism was a religious belief that had no place in the public school science classroom. "In Russia, they teach only one side of things. We call that brainwashing," Lundquist said. "Do you know what our Declaration of Independence says? It says we are endowed by our Creator with certain inalienable rights. That presupposes creationism."[57] But like their counterparts in Palm Beach County, the Broward County conservatives lost.

Meanwhile, gubernatorial candidate Martinez had beaten out his fellow Republicans and faced his Democrat opponent, Steve Pajcic. Evolution was raised as one of the many issues in the race. Martinez felt that the teaching of creationism should be left up to local school districts, while Pajcic said that professional educators should make that call, not politicians.[58] Martinez defeated Pajcic, but he certainly wasn't done with the creationism issue.

"A Heathen Philosophy"

Mere months after winning the election, Martinez and his cabinet members, who served as the state board of education, listened to two men from Escambia County argue against the state's approval of certain science textbooks. This time, Brady, who was now a veteran at speaking before the state board, said he represented a group called the Accuracy in Textbooks Coalition. There were thirteen books on the list for approval that clashed with families' religious beliefs, Brady told the board on February 17, 1987. "You are the only individuals who are able to protect children . . . from being forced, for instance, to give answers on tests that are in direct competition with what they're taught in their churches and their homes, either that or suffer penalties concerning their grades," Brady said. His partner, Philip McDaniel, had a doom-and-gloom message for the board, saying that "the issue here . . . could very well be the survival of this republic as we known [sic] it, if the foundation is eroded by the dogmatic teaching of a heathen philosophy supposedly substantiated by so-called scientific evidence."[59]

Despite sympathizing with the men, Martinez remained true to his opinion that local school boards should be making these types of judgment calls, not the state government. Once the state approves textbooks, school districts must spend at least half of their textbook budget on materials from the approved list. But the remainder of each district's money can be spent on whatever books they wanted. "I agree that I don't believe any youngster out to be taught contradictory to their creed, but that (needs to be supported) by curriculum direction at the district level," Martinez said.[60] Brady was upset, having been stymied by both the state and the school district. He felt that all the people his group represented, whether directly or indirectly, were being left out and discriminated against.

By the end of 1987, Brady found himself facing a different yet related challenge. A high school science teacher in his district had required that his tenth grade students read the Pulitzer Prize–winning book *The Dragons of Eden: Speculations on the Evolution of Human Intelligence,* by

Carl Sagan. A parent who read the book complained that it implied that the Bible is a fairy tale. "I know when I read in a book that God is a myth . . . the lives of children can be destroyed," the mom tearfully told the Escambia County School Board at a November meeting.[61]

The teacher and his principal had met with about twenty concerned parents prior to the board meeting and had worked out a compromise: students who were not comfortable with the book would be allowed to complete alternate assignments. But Brady seized the opportunity to expand the scope of the issue to include the teaching of evolution. Since the high school's administration allowed students to skip the book, since it was perceived to discriminate against religion, Brady attempted to persuade the school board to see evolution in the same light.

His group proposed that a district-wide policy be implemented that directed teachers not to lower students' grades if they refused to answer questions or participate in discussions on human origins. Furthermore, teachers would be required to give twenty-four hours' notice if such subjects were scheduled to come up. The policy should also forbid teachers from making "demeaning, derogatory or condescending" remarks about a student's religious beliefs.[62] Brady threatened to sue if the board turned him down. The board voted unanimously to refer the policy to the superintendent for his consideration.

At the January 1988 board meeting, a decision was announced. Cecil Carlton, assistant superintendent for instruction and curriculum, said that a committee had studied the policy Brady had proposed and determined that it ran counter to state curriculum requirements.[63]

5

"A Conspiracy to Destroy the Faith of Children"

As attempts to water down evolution in the classroom waned in the far northwest corner of the state, the opposite problem flared to life in March 1988 in the far southeast. Unlike the situation in Escambia County, where parents fought against evolution's dominance in the classroom, parents in Broward County tried to stop a teacher from teaching creationism.

Every year when Fort Lauderdale High School biology teacher Burton Atkinson got to the part of the curriculum on evolution, he would balance it with information about creationism. "I will always be under the impression that I must, as a professional, teach both views," he said. "My purpose is to allow the student to make up his mind on how he originated, how the species got here." In his eight years of doing this, Atkinson had never received a complaint. In 1988, though, Atkinson decided not to spend his customary week giving equal time to both evolution and creationism. Instead, he only showed the creationist film *The Fossil Record,* which the Films for Christ Association had produced in 1983. He did that because he felt the biology course overall was heavily "evolution oriented." There was also a test question directing students to "compare the creation and evolution theories." A student in the class told his parents about the movie and test. The family waited until their son was out of the class so as not to risk his facing any repercussions, and then Armand and Gloria Katz told school board member Neil Sterling what had happened. "My feeling is that the weight of scientific evidence favors evolution as the prime theory," Armand Katz said, "but Mr. Atkinson prefers teaching it as if the weight of evidence were equal. His interpretation isn't mine or the scientific community's."[1]

Principal Jacqueline Barber acted swiftly. Atkinson was given verbal and written reprimands, and his lesson plans, classroom materials, and teaching methods were all monitored by the principal and other administration officials. The teacher did admit that he had used an unauthorized movie in his classroom, but even with all the commotion going on over his lessons, his convictions didn't waver. "I feel that there is a bias toward evolution. I had an obligation to share the biases for creation." In his defense, he said that he never explicitly mentioned God or any specific religion, sticking with the general terms "grand omni designer" or "creator." He said, "I didn't have any intent on converting students to my faith."[2]

Even though Atkinson was now under close supervision, the Katzes wanted to be sure such a thing never happened again in the Broward County school district and asked the school board to review the district's policy. Associate Superintendent Nick Fischer responded that evolution was the main scientific theory teachers were required to cover as specified in the district curriculum.[3] Creationism could be mentioned, he said, but it certainly was never meant to be on equal footing with evolution in the classroom.

Angie Matamoros, the county's science curriculum specialist, said, "We're not passing judgment on whether scientific creationism is right or wrong. All we're deciding is whether it can be taught as part of a science class." She also referenced the 1987 U.S. Supreme Court case *Edwards v. Aguillard*, which had made clear that public schools could not mandate the teaching of creationism alongside evolution. That decision about a Louisiana law was a mere nine months in the past. Regardless of the Supreme Court's findings, Atkinson made it clear that the local policy didn't sit well with him. "This is really troubling," he said. "We are essentially muffled as educators to present a viable alternative to evolution."[4]

Atkinson was a member of Citizens for Scientific Integrity (CSI), a group organized in Florida that year that networked with national creationist associations such as the Institute for Creation Research in California. The school district's reaction to Atkinson's mention of creationism in the classroom was a serious issue for CSI members, who felt that society's ills could be attributed to the teaching of evolution. According to CSI board member Ed Hopkins, "If you teach people they're animals, they'll act like animals. They believe they're here by chance and there is no plan for their lives."[5] Hoping to influence the public debate, CSI invited school board members to their Fort Lauderdale headquarters for a viewing of *The Fossil Record*. But due to short notice, no board members attended the screening.

CSI members insisted they were fine with evolution being taught provided that creationism—or "abrupt appearance theory," as they were calling it then—was taught too. This new twist on creationism was developed to counter the Court's determination in *Edwards v. Aguillard* that talking about creating or designing life implies a religious creator or designer. The plan was to simply state that life abruptly appeared without making reference to who or what might have caused that appearance. They said they weren't trying to force their religious views on anyone. "I agree that religion must be kept out of public schools at all costs," CSI member Jim Black said. "But that doesn't alleviate responsibility for putting the evidence into the curriculum. If it happens to agree with the Scriptures, whose fault is that?"[6]

Putting *Pandas and People* in the Classroom

Nearly twenty years had gone by since Rev. Clarence E. Winslow began his campaign to minimize the damage he thought the teaching of evolution was doing to schoolchildren in Florida. Through the mid-1980s he had remained fairly quiet, but in 1989 he once again became a fixture at the Manatee County School Board's meetings. In February he approached the board as the coordinator of the National Task Force for Academic Freedom and said that he was concerned about God having been removed from the classroom. He asked to be placed on the official agenda of a future meeting to discuss the issue. A few months later he made the same plea and finally got a response. The assistant superintendent for instruction would provide Winslow with an update on what was currently in the science curriculum.[7]

Unsatisfied, Winslow returned yet again, this time more passionate than ever. The minutes of the June 6 board meeting reported: "Reverend Winslow stated that he has a signed check in the amount of $1,000.00 payable to any person who can disprove his claim that there is a conspiracy to destroy the faith of children and destroy America through the educational system." He spoke to the board again in July, August, October, and November, each time being denied a place on the official agenda. The school board's attorney said that federal and Supreme Court cases had made it clear that creationism could not be taught in science classes, so there was no reason for the board to get involved.[8]

In January 1990, Winslow returned again and presented the school board with the book *Origins: Creation or Evolution* by Dr. Richard Bliss, published by the Institute for Creation Research. Winslow warned the board that the lawyer for his National Task Force for Academic Freedom was willing to sue if they didn't accept the book for classroom use. He argued that if "viable instructional materials" were available, students had a right to have access to them.[9] The board deferred to their attorney to research the claim.

Winslow's next stop was Pinellas County in March, where he asked the school board to consider putting a new book, *Of Pandas and People: The Central Question of Biological Origins,* in the classrooms. He was turned down. Winslow appeared before the Manatee County School Board again in September 1990 and asked the board to use *Of Pandas and People* as a textbook. The meeting minutes said: "Reverend Winslow stated that if school officials do not comply, there is no choice but to file suit for Constitutional rights of teachers and students."[10]

Intelligent Design

Of Pandas and People was the creationists' answer to their defeat in *Edwards v. Aguillard*. Whereas creationism was essentially barred from mandatory coverage in the public schools' science classes, creationists saw an opportunity in the justices' ruling: "We do not imply that a legislature could never require that scientific critiques of prevailing scientific theories be taught. . . . [T]eaching a variety of scientific theories about the origins of humankind to schoolchildren might be validly done with the clear secular intent of enhancing the effectiveness of science instruction."[11]

Actually, it's too simplistic to say that *Of Pandas and People* was a response to the Court's decision. The Foundation for Thought and Ethics, a nonprofit Christian organization based in Texas, had been working on previous versions of the book a few years prior. Those preliminary texts incorporated the "abrupt appearance" concept coupled with creation references. However, drafts made after the Court's decision attempted to scrub out anything that hinted too much of creationism and replace it with the new terminology of "intelligent design." The strategy was to juxtapose design by an unnamed "intelligent agency" with the theory of evolution. A passage from the book states:

> It is a mistake to claim for macroevolution the status of fact. The existence of fossils with enormous variety is a fact, and so are the changes in the distribution of those fossils over time; to read an evolutionary history of life on earth from the fossils, on the other hand, is to construct a theory. To read intelligent design from the fossils is also to construct a theory. So both Darwinism and design must take their places as theories to be considered and evaluated.[12]

According to the Discovery Institute, a public-policy organization formed in 1990 and headquartered in Seattle, Washington, the concept of intelligent design is totally separate from creationism:

> The theory of intelligent design is simply an effort to empirically detect whether the "apparent design" in nature acknowledged by virtually all biologists is genuine design (the product of an intelligent cause) or is simply the product of an undirected process such as natural selection acting on random variations. . . . Unlike creationism, the scientific theory of intelligent design does not claim that modern biology can identify whether the intelligent cause detected through science is supernatural.[13]

There was another connection between *Of Pandas and People* and Florida besides Winslow's and other's attempts to get it into the classroom. Florida resident Percival Davis co-wrote it with Dean Kenyon, professor emeritus of biology at San Francisco State University. When the book was published, Davis was a professor of life science at Hillsborough Community College in Tampa. A few years later he became a visiting professor of biology at Clearwater Christian College in Clearwater under the name Dr. P. William Davis.

Despite the determined efforts by the book's publishers and promoters to distance intelligent design from religion, they were ultimately unsuccessful. Davis was even quoted in the *Wall Street Journal* about writing the book: "Of course my motives were religious. There's no question about it."[14]

A Higher Authority

Winslow's well-known religious stance tainted his message to the Manatee County School Board that *Of Pandas and People* didn't mention God. He further damaged the cause when he said, "They have locked the creator out of the classroom and allowed atheism and atheistic philosophy to replace belief in God and traditional values." The book was passed along to a textbook evaluation committee, where it was "declined for use as a textbook."[15] The book could be placed in libraries, though, as a resource, they said.

In October the school board listened to Winslow berate them for their choice. According to the minutes, "He stated that the charge against the School Board is a fraudulent misuse of tax funds to destroy faith in the Creator and moral values with tax money."[16] He gave them an ultimatum: either approve in one month the placement *Of Pandas and People* in the classroom, or the National Task Force for Academic Freedom would file suit. In a memo, the school board's attorney explained that the school district simply could not adopt the book.[17]

At the November meeting, Winslow warned that "the time has come to go to a higher authority" and that he "saw no choice but to take it to the Supreme Court." He said that he was "leaving it in the hands of his attorney." Winslow was supported by Sheldon Clements, a parent who told the board that because God had been taken out of the schools there was more drug use, more teen abortions, and more teen suicides. The meeting minutes report: "Mr. Clements stated that if the schools want to teach children that they come from apes, teach it, but teach an alternative, too."[18]

It's unknown if Winslow or his attorney ever followed through with their threats. School board meeting minutes in subsequent months never mention such a lawsuit. Nearly a year went by before the board heard from Winslow again. At a September 1991 meeting he announced he had just received word that the Bible and *Of Pandas and People* were being distributed in the USSR. Furthermore, a recent "Senate Judicial Committee" report claimed that the United States "was the most violent and self-destructive nation on Earth." According to Winslow, this was because of the rise of humanism and the turning away from traditional beliefs and values. He wrapped up his lecture by requesting that students be presented with "alternative views of origins."[19]

Winslow appeared before the board in January 1992, and then again in February. As always, he wanted his issue of origins to be placed on the agenda. At the February meeting he saw a glimmer of hope. Superintendent Gene Witt stressed that the board could not take action on the teaching of creationism due to decisions in federal courts and the U.S. Supreme Court. The board's attorney agreed, pointing out that there had been no new developments concerning the issue and that the school board had no legal authority to do anything even if it wanted to. However, board member Ruby Byrd expressed frustration with Winslow's persistent requests to be put on the agenda and the board's equally persistent denials. Byrd was told that the issue had been thoroughly discussed in the past, before she became a board member. But that response didn't answer her basic question: How long would this request and denial back and forth go on? "Mrs. Byrd stated that she wondered how long the Board is quietly going to sit and listen to Reverend Winslow when he exceeds the three minute time limit, usually speaking for 20 minutes." Why not just put him on the agenda and take care of this for good? Witt replied that the issue had been resolved, but it just wasn't to Winslow's liking and so the reverend refused to give up. Byrd still made a motion to have Winslow placed on a future agenda, but she couldn't get any support to make it happen.[20]

Winslow spoke at a few more board meetings that year, but after that his name was never mentioned again in any school board's meeting minutes or in any news stories. He died in 2002 at the age of ninety-five in Castle Rock, Colorado.[21]

"Some People Believe in the Flat World"

By all accounts, the 1990 gubernatorial race started off as a low-key affair. The incumbent, Bob Martinez, a Republican, was facing off against Lawton Chiles, a Democrat. In September, Martinez, who had taught American history in his early career, held a news conference to outline his future plans for education, which included adding a new state university. Afterward, reporters peppered him with questions, as usual. Revealed during that session was that Martinez believed creationism should be taught alongside evolution in public schools. "There are two viewpoints on it, and . . . certainly you shouldn't deny it or censor it," he said.[22]

That comment made the next day's headlines, but it was hardly a new revelation. Martinez's views on the subject were already public knowledge from the last time he campaigned and from when he was briefly caught up in a stir over textbooks shortly after taking office. After the initial news stories, the subject seemed to fade into the campaign background for a few days. Then, during an impromptu news conference, Chiles and his running mate, Kenneth Hood "Buddy" MacKay, were asked if they agreed with Martinez that creationism should be taught in the schools. "Some people believe in the flat world, I guess," MacKay answered. Suddenly, the once sleepy race for governor burst into flames. MacKay went on to say that he didn't believe in scientific creationism and didn't think it belonged in schools. Instructional time was real tight in the state's classrooms, and he didn't want any distractions added to the day. "I hate to see any more theories put in there," he said.[23]

Chiles seemed to be taken off guard by the subject. He waffled quite a bit, eventually settling on a neutral stance. He didn't think the teaching of creationism should be mandatory, but he felt the issue was best addressed at the local level. "I would not tell them they couldn't do it," he said.[24] MacKay then echoed Chiles's opinion and tried to do some quick damage control, saying that he wasn't equating scientific creationism to belief in a flat earth.

Lawton Chiles (*left*) and Buddy MacKay campaign in Florida's gubernatorial race, 1991. (Phil Pollack, State Archives of Florida.)

But the newspapers had the juicy quote and ran with it, which, of course, then attracted plenty of attention. The organization People for the American Way jumped into the quickly escalating war of words. They felt that both Martinez and Chiles were advocating the teaching of creationism to some degree, despite MacKay's dismissive remark. They sent letters to both candidates' camps asking them to recant their pro-creationism statements. The organization billed itself as a defender of the separation of church and state and called the sudden attention on creationism "pandering to the religious right" that would "sully science education with religion." Neither camp paid the organization much attention. Meanwhile, Martinez went on the attack, claiming that MacKay's "flat earth" remark was an insult to those with strong religious beliefs. "He might as well go ahead and say they should rewrite the Old Testament," he said.[25]

The gubernatorial race was exceptionally tight, with polls indicating the opponents were running even. Martinez leaped at the chance to snatch conservative votes. Less than a week after MacKay's statement, Martinez blasted a radio ad at North Florida and Christian radio stations throughout the state. The thirty-second spot stated: "Lawton Chiles and Buddy MacKay think you share their views. But do you? Recently, Buddy MacKay told the *Tampa Tribune* that anyone who believed in the biblical story of creation would believe the Earth is flat. And when the reporter asked Lawton Chiles, Lawton wouldn't say. So if you believe in creationism, now you know what Chiles-MacKay thinks of you."[26]

In his counterattack, Chiles blamed Martinez for distorting and sensationalizing the issue. He accused his opponent of negative campaigning and twisting his running mate's words in the process. Martinez's campaign manager, J. M. Stipanovich, added fuel to the fire when he shot back: "Why doesn't (Chiles) apologize for having a bigot on the ticket, and maybe I'll think about removing the ad. If Buddy MacKay had said something like that about anybody else's strong beliefs—Jews or Catholics or blacks—he'd be hanging from a nearby tree."[27]

MacKay was stung by the accusation of bigotry. "I have taken some political risks in my career to prevent being called a bigot," he responded. "A lot more than Mac Stipanovich has." Meanwhile, MacKay tried to clarify his creationism stance. His "flat earth" comment was meant as an example of the vast number of beliefs people have, he explained, not all of which deserve to be included in the school curriculum. He also pointed out that Chiles and he, who were both Presbyterians, thought that such issues should be handled at the local level rather than the state level, just as Martinez had stated. Martinez, a Roman Catholic, didn't back off. He stood behind his ad as accurate and claimed that MacKay's statement showed "intolerance for someone else's beliefs."[28]

With the issue at a full boil, newspapers asked Florida's citizens what they thought. The *Orlando Sentinel* asked people to express their opinions in a "Sound Off" call-in poll. Over the course of two days, 5,759 calls poured in, with 57 percent thinking that creationism should be taught in schools. A poll that the *Miami Herald* commissioned from Mason-Dixon Opinion Research Inc. found that 56 percent were in favor of teaching creationism, while only 29 percent opposed it.[29]

The president of the Florida Baptist State Convention, Rev. J. C. Mitchell, jumped into the fray with a statement signed by pastors and other Christian leaders from across the state asking MacKay to apologize: "Surely it is not asking too much of our political leaders, and of those who aspire to such leadership, to refrain from ridiculing the conscientiously held religious beliefs of people within the community of faith." Furthermore, MacKay's comment was "an affront to Christians and Jews who take seriously the biblical account of the origin of the world and the creation of man." Tom Scott, who led the Florida Christian Coalition, added, "This is about religious intolerance. Some 80 percent of the people profess to be religious."[30]

As the pressure mounted, MacKay finally issued an apology. He stressed that he was a "church-going Christian" and firmly believed in freedom of religion for all. "If my comments in Tampa regarding the teaching of creationism in public schools were misunderstood or I did not say them clearly, I apologize to anyone who takes offense." He went on to say, "I apologize if I said something that shows a lack of respect for someone's religious beliefs. . . . The point I was trying to make was the opposite—that we should respect people's views and that government should not impose views on the people."[31]

Of course, creationism was just one issue among many debated during the gubernatorial race. It did attract a lot of attention, though, and put Chiles and MacKay on the defensive for several days. Nonetheless, when election day came, Chiles emerged the victor.

6

"It Was Historic, Wasn't It?"

The Lake County School Board in 1990 was wasting the public's money, local resident Pat Hart believed. There was little discipline in the schools, and no one was teaching the core knowledge students needed for their futures. It got so bad in her mind that she had pulled her son out of public school a few years previous and enrolled him in a private school in a neighboring county. What Hart saw in the schools and her community motivated her to take on two-term incumbent Anna Cowin in the upcoming school board election. Hart's platform included opposition to tax increases and the school district's sex education program. She wanted a return to basics in the classroom and "wholeheartedly" favored prayer in schools. Her campaign literature proclaimed that it was "time to move away from what is new and move toward what has been tried and proven."[1]

The *Orlando Sentinel* endorsed Cowin and expressed reservations about Hart's motivations. Despite praising her hard work and sincerity, it noted that "Ms. Hart's strong fundamentalist approach to issues such as prayer in schools could lead the board into legal quicksand. Moreover, it would appear difficult for voters to support a candidate who has so little faith in the school district she seeks to serve that she sends her teenage son to a private, out-of-county school." The thirty-eight-year-old

Hart's "fundamentalist approach" might have come from her active participation in a local branch of the Concerned Women of America, whose stated mission was "to protect and promote Biblical values among all citizens . . . reversing the decline in moral values in our nation."[2]

Once the votes were counted—and recounted—Hart emerged the victor by twelve votes. The woman from Groveland credited prayer and a strong grass-roots effort for her win. She immediately got to work, publicly protesting a grant the school district was seeking that would establish three campus health clinics. More than a week before her first school board meeting, she was already calling for a halt to the grant process, saying that the clinics might distribute contraceptives or tell students where to get them. Another concern she immediately targeted was the fact that the textbook used in sex education classes leaned heavily in favor of contraceptive use while covering abstinence in only one page. She asked that the textbook be removed.[3]

Hart failed on both efforts. The textbook issue was never seriously considered by the other board members, and hers was the only no vote on the health clinics grant. It was clear that she was alone on the board. "And it appears obvious there will be a lot of 4–1 votes over the next four years," said one observer. Despite her uphill battle, Hart had strong support from conservatives in the community who felt they finally had a champion for their causes on the board. "We have one new board member who reflects a conservative viewpoint and there are people who share those views, but they are a minority," said school board chairman Jerry Smith. "They have been out there all along. I guess now they feel like they have a spokesperson."[4]

Hart labored on. She unsuccessfully challenged the use of children's books that she felt demeaned parents or promoted witchcraft, and she cast the only no vote against the establishment of a before-and-after-school program because she felt it "created another cycle of social welfare."[5]

"Preach the Truth"

Just four months later, Hart was on the winning side of a 3–2 vote that plunged the Lake County School Board and the community into chaos. The subject: creationism. During the March 26, 1991, board meeting, citizen Bob Wells asked the board for permission to go into the schools and teach "the subject of creation according to the scripture, the Bible." Wells had obtained his ministerial certificate in a home-study Bible program and was director of the Lake County Christian Care Center, which provided food and other materials for the needy. It was pointed out that since his request was not on the agenda, the board members could not take action. Wells subsequently submitted a handwritten note to request time on a future agenda: "I make a request that the Genesis account of creation be taught in the Lake County public schools."[6]

A few days later, Linda Rozar, president of Concerned Citizens of Florida and the Lake County American Family Association, sent a letter to the board:

> Citizens within the county have raised the question of teaching creationism in the public schools. Their concerns focus on the ever expanding scientific knowledge pointing to intellegent [sic] design as opposed to evolving process of life.
>
> It is our hope that the board would concider [sic] workshops to review materials from this perspective and add a creation unit of study to the science curriculum used in the various grades.
>
> Please contact us if we can be of assistance regarding texts, student workbooks, videos, etc. Thank you.[7]

Rozar lamented the removal of prayer and the Ten Commandments from public schools, claiming that "true Christian freedoms" were eroding. Likewise, Wells felt that students needed an awareness of "the Great Creator" in order to stem the growing tide of teen pregnancy and school dropouts. The fifty-two-year-old Wells hoped that creationism lessons offered on a voluntary basis would be a start. "When they took (the) Bible and prayer out of schools, other things began to take its place," Wells said.[8]

Wells's humble, single-sentence request ignited a community crusade that overwhelmed school board meetings with hours of debate. Board members seemed ill prepared for the more than 150 citizens who crowded into the first meeting addressing the matter, on April 23, 1991.[9] There were so many in attendance that an overflow crowd had to watch the proceedings on closed-circuit TV in the lobby. Chairman Smith initially advised everyone that only thirty minutes was typically allotted for each agenda item.[10] However, the discussion stretched out for two hours with about twenty-five citizens addressing the board.

Wells was the first to speak. He spent a little more than more ten minutes voicing his concerns about teen pregnancy, abortion, and violence in schools. He remembered that when he was a kid in school there were prayers and scripture reading during the day that instilled in the kids a sense of respect and reverence. He then talked about the need to teach creationism alongside evolution in the interest of intellectual honesty. The Supreme Court had determined that teachers possessed the flexibility to supplement science instruction, he claimed. The audience punctuated his speech with applause and amens throughout.[11]

Wells then illustrated his concerns with a scenario:

> Suppose little Johnny goes to school. He hears evolution and that's all he's fed, a diet of evolution, which is not a fact, it is only a theory. . . . In the school he hears evolution . . . in some of them they'll say the Bible is not valid. It is not credible. All right, little Johnny goes to the church. He hears the pastor say evolution is not valid. Evolution is not credible. The Bible is right and that's all. Little Johnny, he don't know who to believe. He don't know what to believe. He's so confused now that he loses credibility in the schools and possibly in the churches and then he's left as a casualty. I'll say it again as solid as I can: if we're going to teach one theory in the school we ought to include the other along with it.[12]

Such confusion and distrust fostered in children is what leads to a nation in trouble, he warned. To turn this around, Wells demanded that the home, the church, and the school all join hands to give children stronger moral standards. One way to do that was "to give them the best balanced education we can give them," he said.[13] The crowd erupted into loud and long applause as he finished.

Chairman Smith then called on school board attorney Walt McLin, who had studied the issue in preparation for this meeting. But McLin pointed out that it looked like others in the audience had more to say. When Smith asked if anyone wanted to come forward, several hands went up. Rozar's was the first. Rozar asked that the board members conduct a workshop about including creationism in the curriculum. She gave them a multi-page handout and asked them to read through a list of what she said were problems with the theory of evolution. Additionally, including the Bible in such discussions shouldn't be a problem, since the Supreme Court in 1782 had approved its use in schools, she erroneously claimed. Even teachers who might be uncomfortable with using the Bible in class can rest assured that reading from the first chapter is not proselytizing. "Genesis 1 is an order of creation," Rozar said. "We ask that order to be taught. Genesis 1 will not convert a person to Christianity. Genesis 1 cannot be used by any religion for converting a person."[14]

From there a long parade of citizens spoke their minds on the issue. Only a couple of people opposed inclusion of creationism, and a couple of other speakers' statements were hard to categorize. Everyone else sided with Wells and even echoed his motivations. Many people expressed concern about society's declining morals and disrespectful children. No one went so far as to directly blame society's ills on the teaching of evolution—though some hinted at it—but they all wanted some form of religion in schools. "I feel that God is good for our students," one man said. "I feel that lack of respect for God is bad for our students." "We've got a confused generation," an older man said. "We need to get back on Biblical grounds. We need to preach the truth."[15]

A former junior high school teacher reflected on his time in a tough classroom: "I had a paddle in one pocket and a Bible in the other. That's back when we could paddle children when they needed it." He said that he was an ordained minister and started a Bible club at the school in an effort to tame the out-of-control students. "The children were selling themselves in the portables when I got there. You know what they were giving the girls? A dollar a throw they called it, during lunchtime. We didn't have that after we had a Bible club started."[16]

Many said that there was no evidence for evolution and that the available evidence fit creationism better. Evolution is a faith and a religion, they claimed. "Here' a little quick quiz," one man said. "What hatches from a chicken's egg? This is multiple-choice, three answers: a cat, a dog, or a chicken. No one has ever seen anything hatch from the egg of a chicken other than a chicken. There's more scientific evidence that supports Santa Claus than there is that supports evolution."[17]

"Creationism is a good science," another man said. "It fits the fossil record. It fits the first and second and the . . . third law of thermodynamics.

It fits into the realm of science as well as or better than, in my opinion, the theory of evolution. I believe creation should be taught because it is good science. The second reason I think creation should be taught is because evolution is a faith, and if one faith is taught then fairness and equity only require the fact that the second faith be taught. Evolution is a faith. There's no evidence for it, no repeatable evidence and it's taken by faith because no one was there to observe it."[18]

Those who opposed the teaching of creationism were clearly in the minority that night, with only two men stepping up to speak. The first read a statement from his local Unitarian group: "The majority of Lake County taxpayers would refuse their tax money being spent to indoctrinate their children with one particular religious view posed under the guise of science. It is certainly offensive to members of this Unitarian group when people pigeonhole an entire belief structure according to their own biases." The other man went a step further, saying that just because a majority of those present at the meeting were Christians demanding that creationism be taught in the schools, that didn't mean they spoke for all Christians. "Well, I'm a Christian and I don't want creationism taught in the schools."[19]

"Our Children Are Not Bad"

Finally, after more than an hour of public testimony, the chairman called upon attorney McLin to offer his advice. "This is one of those unenviable tasks" in which you are the bearer of bad news, McLin said.[20] First, he pointed out that the agenda item before the board clearly asked for the Genesis account of creation to be taught in the county's schools. It "unequivocally" can't, he said. It's the establishment of a religion, which violates the First Amendment of the Constitution as proven several times in federal courts and the U.S. Supreme Court. McLin acknowledged that the evening's speakers had moved beyond the original agenda item by requesting that evolution and creationism be taught together. Court cases had already addressed that exact proposal, and it can't be done, he said. The courts had left open the possibility of creationism being taught in a comparative religion course, but not in a science course.

Wells was then allowed to respond. "You said part of the truth but not all of the truth," he said. "The Supreme Court noted that teachers already possess the flexibility to supplement the present science curriculum with the presentation of theories besides evolution about the origin of life and are free to teach all facets of this subject of all scientific theories about the origins of human kind. They are free to do it." A couple of other citizens were concerned that there appeared to be conflicting opinions and information. A man pointed out that in California "both are required to be taught" and that in Alabama "the law says that all theories of the origin of life shall be taught."[21]

Board member Hart agreed that there was a lot of confusion and that a workshop was needed. She even added to the confusion with her own bombshell announcement: "I have been in contact with the DOE [Florida Department of Education] in Tallahassee and they have said there is nothing to preclude us from teaching creationism. And if you go and look at our county curriculum guidelines and also the state curriculum guidelines it just says biological change through time. There's no precluding of us teaching creationism in our own guidelines."[22] When asked whom she had talked with in the capital, she identified state science specialist Dick Tillis.

Other board members joined in the discussion. Sandra Green fired back at many of the evening's speakers. "I do take objection when people in this audience say our children are bad," she said with passion obvious in her voice. "Our children are not bad. We have some very good students in our schools in Lake County. They are very highly educated, very highly motivated. They are dedicated to being good, morally responsible young people in this county. And I do take offense to people that say these kids are otherwise."[23] She went on to say that it was the responsibility of the schools to educate, but it was up to homes to handle the training for such things as morality.

"I'm hearing a different message, obviously, than Mrs. Green," Hart countered. "I'm not hearing that we're asking people to teach our children to believe. We want them to have the opportunity to study, to understand. Mrs. Green is right. It is the role of the home and the church to teach to believe. It is the role of the school to study to understand. If they are not offered both of the different models, how can we teach them to understand?"[24]

More discussion ensued with board members and citizens all quickly making various points that left board member Phyllis Patten exasperated. "I don't understand many of the things you people are saying to me," she said as she agreed that a workshop might be helpful. Chairman Smith tried to shepherd the discussion toward a conclusion, because Hart had made a motion to schedule a workshop that the board needed to vote on. He stated that he was against the workshop because voting for it would give citizens "false hope" when in fact there was no way the board would take the financial risk of going to court to defend the teaching of creationism, especially when the school system would clearly lose.[25] Board member Tim Sullivan agreed with him, while Superintendent Tom Sanders declined to give an opinion. Despite Smith's efforts to get the voting over with, there was still quite a bit of haggling over what the workshop entailed.

Once it was finally clarified that the workshop was strictly for information gathering and not for any definitive action, the vote was taken. Hart, Green, and Patten voted for it, and Smith and Sullivan voted against. Applause broke out in celebration of the victory, but when it faded Green admonished the crowd that there would be strict guidelines to follow: "The religion part of this must be left out. Now I don't know how you can do that, because my religion doesn't let me do that. No, mine doesn't. O.K. So that eliminates three quarters of the people who turned out tonight." Patten explained that she had voted for the workshop because with all of the conflicting and confusing information presented the previous two hours, those who attended deserved clarification on what could and could not be done. "Some of us may walk away not happy, but at least we will have invested the time."[26]

The board discussed when to have the workshop and settled on May 13 at 7 p.m. Smith tried to move on to the next item on the agenda, but then a woman asked how the workshop would be run. Would everyone be allowed to speak? Would there be sign-up sheets? Smith explained that workshops were usually informal and that everyone could participate, but then other board members interjected to demand more structure, such as time limits on speakers and the elimination of repetitiveness. Hart wanted to bring in experts to present information. Green was very cautious, saying: "We can get opinions all night, people. And it's not going to do us any good."[27] It was the question of legality that needed to be resolved. Finally, it was determined that there would be three speakers for each side of the issue. Each would speak for five minutes, and then all others wishing to speak would be limited to three minutes.

After the meeting, Patten told reporters: "I wasn't sure what they wanted. I am not certain they were sure what they wanted. It is an emotional issue, but you can't teach based on the emotions." The next day Smith said: "It was historic, wasn't it? What can I say? I guess it's one of those things where persistence pays off."[28]

"Religious Evolution"

The next three weeks saw a flurry of activity as groups organized and newspapers ran a string of guest opinion columns. Of particular significance was the *Orlando Sentinel*'s interview with Tillis, who Hart said had told her teaching creationism would be permissible.

> Tillis said Wednesday his comments had been misconstrued by Hart.
>
> The state's guide to science instruction says any subject that relates to a "biological change through time" can be offered, Tillis said. Hart might have taken that to mean any subject believed to be scientific could be offered, Tillis said.
>
> "What I told her was there was nothing that the state department sends out that tells you what you can and cannot teach," he said. "She asked if there were school districts that teach it, and I knew of no such thing."[29]

Likewise, the Lake County newspaper, the *Daily Commercial*, interviewed Robert Lumsden, who was in charge of math and science programs in the state department of education's curriculum department. He confirmed that no one at the department of education had told Hart that teaching creationism was acceptable. "There are those who feel that curriculum ought to include astrology or pyramidology or almost anything else," he told the newspaper. Despite these revelations, Hart told the *Orlando Sentinel* that she stood by her prior statements. The *Sentinel* also invited citizens to participate in a "Sound Off" call-in poll. Over the course of two days, 4,356 voiced their opinions, with the majority (2,591) supporting the inclusion of scientific creationism in the public school science curriculum.[30]

As word of the upcoming workshop spread, Lake County suddenly found itself the focus of national attention. John Morris, representing the Institute for Creation Research in California, told the Florida media that evolution and creationism are both religious views. "For the evolutionist to say science has to be a natural process that excludes the supernatural is a religious statement," he said. "If you're going to have one religion in the classroom, you'd better have the other." He added that the creationism push came from "people being sick and tired of having the government use tax money to shove religious evolution down the throats of students."[31]

The pro-evolution organization National Center for Science Education, also from California, offered an opposing opinion. Executive Director Eugenie Scott said, "Scientific creationism is a biblical (teaching) disguised by science—it's outside the scientific realm because it allows the intrusion of the supernatural." Freedom of speech only goes so far, she added. "Should we teach that AIDS is a curse from God, or that AIDS is caused by a virus? We don't teach ideas that haven't stood the test of time and we don't teach what pressure groups think we should teach." Those in favor of teaching creationism were vocal and organized, but the opposition wasn't far behind. The board of directors of the local teachers union voted unanimously to take a stand. "We believe it would be a fundamental mistake to provide for the teaching of scientific creationism in the science classes of our schools," their press release stated.[32]

A couple of days later, several more organizations threatened a lawsuit and election-time retribution toward school board members if creationism made it into the science classroom. Joining forces were the statewide teachers union Florida Education Association–United (FEA-United), the ACLU, the American Jewish Congress, the National Conference of Christians and Jews, the Anti-Defamation League of B'nai B'rith, Americans United for Separation of Church and State, Florida's Association of District School Superintendents, the state Association of School Administrators, and the state PTA. "The state teachers union would take the lead in bringing a lawsuit against the School Board if they do not reject the proposal," FEA-United spokesman Ron Sachs told the *Orlando Sentinel*. "By keeping this proposal alive, school officials are pandering to a fundamental religious group." He added, "I can't believe it has gotten this far."[33]

The controversy spawned the creation of homegrown groups. One citizen organization, Concerned Lake County Taxpayers for Separation of Church and State, focused on the money that would be wasted if a court case became reality. Their opposition was the Coalition for the Secular Theory of Special Creation, whose members coalesced around the issue of censorship. They demanded "intellectual honesty, truth and fairness." Speaking for this group was one of its founders, Coy Jamerson III. "We wish to make it clear that science teachers have the right to teach science facts and students have the right to hear them," he said. "Our real enemies here are those who wish to continue to censor scientific law and facts of the only other valid scientific theory of origins, the secular theory of creation."[34]

Working diligently for the creationism side was Grady McMurtry, president of Orlando-based Creation Worldview Ministries, an "educational missionary organization" he created in the early 1990s. McMurtry attempted to recruit the National Association for the Advancement of Colored People by sending the following letter to the local and state president:

> To T. H. Poole Sr., from supporters of the creation theory:
> I have enclosed two pages which clearly show that . . . the evolutionary theory proved white superiority. Creationists are strongly anti-racist.
> Please contact me at my home or office for more information on social Darwinism or related subjects.[35]

His appeal didn't work. "I take offense when this is used as a prodding tool," Poole said.[36]

The *Orlando Sentinel* asked each school board member what his or her thoughts were on the teaching of creationism. It became clear that despite the 3–2 vote favoring the workshop, a 4–1 vote defeating creationism was likely. Chairman Smith said that scientific creationism was "a religious theory I happen to believe in," but he stood firm against taking any unconstitutional actions. Patten was also confident: "It hasn't been proven through a scientific procedure. It's a religion based on belief." Sullivan didn't seem to dismiss scientific creationism outright as Patten did, saying: "It has scientific merit, but like anything that has come through history, fact and legend have become intertwined." It didn't belong in the science class, though. "It is something ministers need to be teaching in Sunday school."[37]

Stand Up to "Outsiders"

Finally, the public workshop kicked off on May 13, a Monday night, at the Eustis High School gymnasium. Newspapers had varying crowd estimates, with the *Daily Commercial* saying about 400 while the *Orlando Sentinel* put the number at more than 550. Regardless of the size, the crowd was polarized and energized for a raucous debate.

Just getting the event started led to conflict. The original plan was to allow three speakers from each side of the issue to talk for five minutes. However, school board members received a lot of complaints prior to the meeting, so the first topic up for discussion was modifying that rule. Smith offered to his fellow board members a proposal of twenty minutes for each of the initial six presenters, but Patten felt that ten minutes would be good enough, and Sullivan agreed. It looked like the issue would be settled when Hart then asked for a compromise of fifteen minutes. Other board members quickly shut her down, saying that the ten-minute proposal was already a compromise. A vote was held to accept the new time limit, and all approved except for Hart.[38]

Wells stepped up to a microphone and complained that ten minutes wouldn't be enough time. Smith told him that the issue had already been decided and that the clock was now running. Wells didn't speak, though; instead, he introduced the first speaker for the creationism side: Stephen Hurm. He was president of the Rutherford Institute of Florida, a religious liberties group, and he was a former local prosecutor. Hurm emphasized that no one wanted to mandate anything in the schools. Instead, they were rallying against censorship. The purpose of education is to spread views, not stifle them, he said. The people he represented wanted teachers to be able to use supplemental instructional materials if they wished to do so. To justify their stance, Hurm quoted the district's own policy on controversial subjects: "Controversial material may be presented if it's germane to and relevant to the subject area being taught."[39] He argued that origin theories other than evolution fell into that category.

Then Hurm referenced the 1987 Louisiana evolution case *Edwards v.*

Aguillard that went to the U.S. Supreme Court, explaining that the justices had said teachers could present scientific theories other than evolution. No law prohibits it, he said. Hurm concluded with a request that the board recognize that teachers already had the flexibility to supplement biology instruction and that the board allow supplemental materials be presented to the curriculum committee for use in the classrooms. The crowd gave Hurm a long and loud standing ovation. Once it was over, Smith asked the audience to refrain from such outbursts again, since they took time from the already-packed schedule and "detract[ed] from the proceedings."[40]

The next two speakers were Grady McMurtry and Dr. Richard Lumsden. McMurtry billed himself as a former atheist scientist, and he was a member of several creationist organizations, including the Creation Research Society. He lectured the school board about evolution being unproven, the universe being too young for evolution to happen, and problems with the fossil record. Lumsden also was known as an atheist scientist who investigated evolution later in life and found it lacking. He became a Christian creationist and traveled around the country to argue against evolution. His points to the board were similar to McMurtry's: evolution is a woefully inadequate explanation for life's origins and diversity, is based on faith, and contradicts natural laws as we know them.[41]

Whereas the creationist speakers appeared to have coordinated before the workshop, the pro-evolution side seemed to lack that level of organization. There was momentary confusion as the chairman asked for three speakers to identify themselves as being present. Local attorney Jerri Blair was the first to speak. He stated that he was there on behalf of the group Concerned Taxpayers for Separation of Church and State, which was formed with the goal of staving off lawsuits over the current matter. There could be no doubt that litigation was in the future if the board chose to include creationism, he warned. Blair's two main points were that the school board was being asked to do something blatantly illegal and that if they did so, taxpayers' money would be spent on litigation rather than students.[42]

Blair's remarks didn't sit well with the crowd. Only a minute into his presentation, audience members began to voice their disagreement, forcing the board chairman to rap his gavel and scold them. Next, Blair argued that the Supreme Court had clearly defined creationism as a religious view and that those views could not be forced on students. Students are required to attend school, so they are a captive audience for whatever message that is given in the classroom. Parents trust their children to the school system. Any religious view officially endorsed in class could be different from the religious beliefs practiced at home and thus violate that trust. Blair finished before his time was up, asking the board to heed sound legal advice to not allow creationism to be taught in the science classroom.[43]

Next to speak was Dr. Linda Wolfe, an anthropology professor at the University of Florida in Gainesville. She explained that her specialty was the research of human evolution and the natural behavior of wild monkeys, which sparked some commotion and laughter from audience members. She pressed on, saying that the teaching of evolution was bearing the brunt of the blame for society's ills by those wanting creationism to be taught in the schools. But they are clearly wrong, she said: "In the past eleven years of teaching it's a rare occasion when I have a student who has the vaguest notion of what evolution is all about. In my experience, evolution is not being taught in Florida's public schools in any recognizable form."[44] She went on to say that creationism was based on a supernatural creator and had no place in a science classroom.

The third pro-evolution speaker, Stephen Judson, another local attorney, took only a few minutes to say that the creationism issue should be a very low priority for the school board. Funding education overall and planning for future student population growth were the real important issues.[45]

Since there was still plenty of time left on the clock, Gail Burry, president of the Lake County Education Association, was allowed to speak. She acknowledged that there were strong emotional reactions to this issue but that the focus should be on preparing children for the future, not imposing religious beliefs. She read a list of organizations opposed to the teaching of creationism, and when she got to the ACLU there was a loud negative reaction from the audience.[46]

Warren Keiner, chairman of the central Florida chapter of the ACLU, then wrapped things up for the anti-creationism side. He said he felt like Rip Van Winkle. It was as if he was in 1925 during the famous Scopes "Monkey Trial" in Tennessee. After a brief lecture on the history of court cases addressing the teaching of creationism in the classroom, he said, "I'm not in 1925 or 1968. I'm in Eustis addressing the same issue that had already been decided."[47]

The chairman then opened the floor to any citizen who wished to speak. For the next hour and a half the board listened to about three dozen people offer their opinions and advice. There were roughly twice as many creationism supporters as those in opposition. The crowd was lively, with regular rounds of applause for nearly every speaker, and several disruptive commotions forced the chairman to pound his gavel and reprimand the crowd. Lawyers in general and the ACLU in particular were the butts of several jokes. Many speakers were concerned about God being removed from schools. Some stated that evolution was based on solid science, while others countered that it was "just a theory."

Some people wanted to avoid the expense of a lawsuit, but they were countered by speakers who offered to help foot the bill. One pastor encouraged the board members to stand up to "outsiders" threatening lawsuits and not let them bully the board into voting against their beliefs. "The people of Lake County elected you to be their representatives, and we are here tonight to give you the moral support you need," he said.[48] He offered to help financially even if the issue went all the way to the Supreme Court.

There were comments about censorship, humanism, the fossil record, the "religious right," the occult, giving students choices, teen pregnancy, integrity, petitions, freedom of speech, the founding fathers, drugs, the Grand Canyon, and even Iran.

"You're Indoctrinating Them in Evolution"

Finally, public comment was cut off and it was time for the board members to talk. No vote would be taken that night, but the members were free to discuss and debate the subject. Patten started by pointing out how the original request that sparked the whole controversy was Wells's asking that the Genesis account be taught in the schools. She said that as the request changed, she studied all she could about what was being proposed. It was clear to her that teaching creationism would be against the law. "I've personally been thinking about this and worrying about this and praying about this and reading about this and I will tell you something," Patten said. "Religion, my religion, is based on belief and a spiritual system, and when I think that we are trying to make religion a science, I cringe."[49]

Greene spoke next. She pointed out that the opinion of the school board's attorney was that teaching creationism would be illegal. But then Hart had spoken with Tillis at the state department of education, who said it would be OK. Greene explained that that was why she agreed to have the workshop. However, some board members have a special agenda, she said. She called Tillis herself to hear it straight from him. "And his first comment to me was, 'Mrs. Greene, my first comment to Mrs. Hart was that it is a legal matter and she needs to discuss it with her school board attorney,'" she said. She went on to say, "I resent when people blame the schools for something that is the family's responsibility."[50] Schools teach students to read, which then allows them to decide the truth for themselves. They will be able to read the most important book every written: the Bible.

Hart was now on the defensive. She said that she had indeed spoken with Tillis and had written down everything he had told her. He had stated that the state curriculum guidelines don't preclude the teaching of creationism, and Hart had no idea why he would change his mind upon talking to Greene. Hart pointed out that the county's science curriculum team's written philosophy says that one thing they would like to accomplish is to develop an investigative and questioning nature to assist in the evaluation of the world. "As far as sitting and indoctrinating children in the classroom . . . you're indoctrinating them in evolution," she said. "Scientific creationism does stand up under the scientific criteria just as evolution. Why not teach them both?"[51] She then pointed out that she had no agenda, as others were insinuating. Rather, a citizen had brought up creationism, not her. When she was finished, the crowd responded with loud and long applause, forcing the chairman to yet again pound his gavel.

School board attorney McLin then provided a long summary of the current Lake County situation along with a recitation of several federal court cases that addressed the issue. He pointed out that one of the creationism speakers that night, McMurtry, was a member of the Creation Research Society, an organization that was mentioned in the 1987 Supreme Court case that found the teaching of creation science unconstitutional. That case had clearly pointed out the religious nature of the group: all members were required to agree that the Bible is the written word of God and that, because it is inspired throughout, all of its assertions are historically and scientifically true. McLin explained that the Court used this information in its final determination that schools cannot teach religious doctrine disguised as science. In McLin's professional opinion, there was no legal way to teach creation science in conjunction with evolution.[52]

The meeting wrapped up with a short debate between Hart and McLin. Hart explained that the board's policy concerning science education made it clear that knowledge should be made available without bias. Then she referred to the controversial materials policy that Hurm had quoted from earlier in the evening. She challenged McLin, saying that he was claiming there was no policy that supported the teaching of creationism, while she had just found plenty of such support. McLin's response was that the cases he had already summarized had found that tactic to be a guise to mandate the teaching of a religion in the classroom. There is no request for the teaching of creationism to be mandated, Hart countered. There was just a plea to allow the use of supplemental material. This sparked applause and a "That's right" from the audience. Greene then stepped in, saying that the policy was being stretched to fit an agenda using a backdoor approach.[53] There was another, if shorter, burst of applause.

Smith took a break in the action as an opportunity to bring the long meeting to a close.

"Preconceived, Prejudicial Judgment"

The very next evening, the school board had its regularly scheduled meeting. A decision on creationism was the first item on the agenda. Citizen attendance was light, with only about two dozen people on hand.[54]

Throughout the previous several weeks, Superintendent Sanders had been quiet, only offering a few quick observations and then deferring to the school board's attorney. However, that didn't mean he wasn't engaged in the issue. At this meeting he took several minutes to recap what had been going on, explaining that those on both sides of the issue were sincere in their stances. As the situation had unfolded, Sanders said, he had consulted with many different people. One thing that had become evident to him was that if the board approved the teaching of creationism, it would be difficult to figure out a way to do it. He discovered that there were many different versions as to what could be taught, and varying ideas as to how it ought to be incorporated into classroom lessons. He spoke with ministers who clearly believed in creationism but didn't believe the school system could teach it correctly. They told Sanders that such matters should stay in the church. Sanders's recommendation to the board was to heed the school board attorney's advice. However, he said that creationist materials could be submitted for inclusion in school libraries provided they were subjected to the regular approval process.

Board members spent a few minutes discussing what could and could not be included in libraries. Could a teacher make the study of those materials mandatory? No. Where would the items come from? Donations. Would schools be required to purchase some materials? No. Would some materials be banned? Anything could be submitted for consideration, but it would have to go through the standard review and approval process already in place.[55]

Once everyone seemed satisfied with the libraries issue, Hart snapped the focus back on the teaching of creationism by attacking the superintendent. She asked if he was recommending that theories of origins can be taught but not scientific creationism. Sanders corrected her, saying that he was merely recommending that creationism not be taught. Hart shot back that this was just a play on words. "Theories can be taught, but not the theory of scientific creationism," she said. "That's all I needed to clarify. Thank you."[56]

Chairman Smith defended the superintendent. "That's not my understanding of the discussion," he said. "The discussion centers on: is the school board going to take a position on shall creationism be taught in the science curriculum." Later, he followed up, saying, "I certainly don't see this as a discussion on theories of origin, what will and will not be discussed. That might be a topic for another date, or another school board in another county."[57]

The public was then permitted to offer comments. Among them was Rozar, who apparently realized that the creationism effort was doomed and attempted some damage control. She said that the workshop was supposed to have been about evidence but wound up being just about drama. Rozar was concerned about students who would have their beliefs violated by requirements to answer questions or participate in activities about natural selection. Their grades would be a stake, so couldn't they be allowed to simply not participate or be allowed to do their own research on other origins theories? Her hope was that students could be free to present in the classroom their own science data learned at home when it differed from the teacher's. She then suggested that the board not take any action and instead table the issue for another time.[58]

Another citizen, Kevin Cottrell, then grilled the superintendent and school board attorney. His concern was that the school board was on the verge of censorship. He asked if teachers could bring up creationism during a general classroom conversation or if it would be illegal for a student to bring it up. Both McLin and Sanders explained that they couldn't possibly answer every "What if" question that citizens might pose. The bottom line was that teachers could not intentionally introduce creationism into a lesson. McLin assured Cottrell that there were no hidden agendas in the superintendent's motion to the board.[59]

Next up was the man whose one-sentence request had lit the community firestorm. Wells was angry and frustrated. The presenters on behalf of creationism at the workshop had given good, convincing presentations about the legal right for creationism to be included in the science curriculum, he said, but it was clear that board members had made up their minds before hearing and examining all the facts. "To me that is preconceived, prejudicial judgment," Wells complained. "That is not being open-minded and fair. That is not the American way. That is not being intellectually honest." In an act of protest, Wells announced that he was going to walk out and have no further part in the discussion. This announcement was met with some weak applause, completely unlike the frequent, enthusiastic outbursts at the previous night's workshop. The chairman then announced that there would be a momentary pause to allow Wells and anyone who would like to join him to depart.[60]

Before the meeting there had been rumors that Stephen Hurm from the Rutherford Institute would bring a lawsuit against the school board if it voted against creationism. However, Hurm wasn't present at the Tuesday night meeting. Coy Jamerson of the Coalition for the Secular

Theory of Creationism told the board: "I would like to just say that that did not come from this coalition. That's not the way we work. We don't care to drag the taxpayers of Lake County into a lawsuit." Jamerson also expressed appreciation to the board for hosting the workshop and considering the inclusion of supplemental materials in the schools' media centers. "Our whole goal here is education," he said. "And we want to continue to inform individuals that this is not a religious issue, that there is scientific evidence that we want to have people to be able to see so that they can learn and make decisions for themselves."[61]

Finally, the chairman said he was about to call for a vote on the superintendent's recommendation, but first he cautioned his fellow board members: "Failure to support this would possibly indicate our unwillingness to work on a very serious issue that a segment of the community has brought to our attention. And at the same time it's my own personal belief that it would show a disregard for the law of the land and I would encourage us to carefully consider as we vote at this time."[62]

The final vote was unsurprising: four votes for turning down creationism and one vote, Hart's, dissenting.[63]

Despite saying that he had washed his hands of controversy, Wells was back in the news a couple of weeks later. He created a nonprofit foundation called Active Christians to Serve. The group's focus was on holding seminars and workshops about scientific creationism. "They think the creationism issue is dead, but it's not," Wells said. "We'll show that when Christians work together for a cause, it becomes a reality."[64] It's unknown if the group ever actually did anything, though, since it was never mentioned in the news again.

"One of the Primal Evils in Our Country"

Efforts to promote creationism in Volusia County had been ongoing for years, but they had never made a big splash. Advocates were no less passionate there than in other counties, but they were stymied at every turn. They felt their views were ignored by the public school system while secular humanism was give preferential treatment. Resident Dick Smith said in 1991, "The losers are the kids. They are being brainwashed."[1]

The Volusia County creationists had worked on state legislators and school board members in an effort to be heard, but without luck. Then they talked directly with teachers and school system administrators in an effort to show them the supporting evidence for creationism. They lamented the fact that the school district used textbooks that either never mention creationism or mention it in a negative light. However, their efforts were constantly frustrated. The district's science specialist, Billie Wisniewski, was matter-of-fact about the issue: "It has not, really, a place in a science textbook. It's not a science."[2]

Creationism was one of several points of contention in a growing conflict between Volusia County–based Stetson University and one of Stetson's benefactors, the Florida Baptist Convention. The convention had given the school more than a million dollars annually in the late 1980s, but tensions grew between the private school and the religious organization. Their relationship started in 1885, but more than a century later the university wanted more control over its own affairs. At the same time, the convention had grown more conservative and disagreed with many things the school was allowing. Control of the convention was turning over to fundamentalists who felt they should have a louder voice in Baptist schools' academic programs.

The convention leadership didn't like Stetson's liberal and historical take on the Bible, its omission of creationism from the curriculum, and reports that condoms had been handed out on campus by a student group. In the early 1990s an agreement was struck that the university would have greater academic freedom and more control over selecting its own board of trustees, while the convention would decrease its funding. By 1993 all funding had stopped. Stetson's president at the time, Douglas Lee, had helped broker the deal in the name of academic freedom. The convention has "become so narrowly focused on theological issues that we, as an educational institution that fosters diversity and an open search for the truth, cannot continue (the same relationship)," he said.[3]

"An Educated Guess"

Anti-evolution sentiments had certainly made big splashes in Manatee County over the years. Even though no one made lasting headway into challenging the teaching of evolution in schools, the feeling that schools were "shoving evolution" down children's throats still bothered many residents in the early 1990s. The Manatee County Ministerial Association decided to take action in 1991 by raising money to purchase copies of the book *Of Pandas and People*. The books would then be donated to public school libraries. Association chairman Rev. Cornell Haan explained, "Schools are just teaching one theory. Neither one is proven, they are both still theories. Both should be taught in school."[4]

The county's science supervisor, Dr. John Bernreuter, said it would be up to the appropriate review committees whether the books made it into libraries, but they definitely wouldn't appear in science classrooms. "The book presented views in a religious manner, which is not appropriate for science," he said.[5] Bernreuter started his job right at the tail end of the last wave of creationism efforts, but he was already confident in his stance. Evolution was taught in the schools, he confirmed, and it was an important concept for students to understand. Creationism isn't a science. "It's based on faith and it's difficult to teach that in science," he said. "Theory is based on evidence. Where's the evidence that there's somebody moving things around? That's faith."[6]

There were those in the school system who disagreed with Bernreuter, though. The chairman of the science department at Palmetto High School, James Pauley, said that ever since the Supreme Court had banned prayer in schools, things started going downhill. He had taught honors biology, which includes a section on evolution. At the end of the unit the students staged a debate on the subject, which Pauley tried to stay neutral on. But he went on to say that if students "saw me outside of school they'd know how I felt on that." Pauley felt that there was no reason to exclude creationism from the schools. "There are lots of theories for many things," he said. "Theory is nothing more than an educated guess."[7]

"Insecurity, Arrogance, and Censorship"

The debate wasn't restricted to the high school classroom, though. In January 1992, creationist Duane Gish from the Institute for Creation Research in California made an appearance in Manatee County. He faced off against University of South Florida anthropologist Lorena Madrigal at the Manatee Convention and Civic Center. Tickets were only a few dollars each, but free tickets made it into the hands of Manatee County School Board members and other key personnel. "Someone made sure I got a ticket to that," Bernreuter said.[8]

The event's organizers claimed that the debate would be all about science and education, not philosophy and religion. Reflecting back on it in 2011, Madrigal didn't remember it that way. In the nearly two decades since the event, she hadn't engaged in any other evolution-versus-creationism debates because of the experience. She had prepared for the debate as if she was going to give an educational public lecture to a receptive audience. She wasn't familiar with Gish, and the Internet was still in its infancy, so finding information about him was difficult.

Clues about how the evening would go started popping up, though. A draft copy of a flyer promoting the event was given to Madrigal for approval. It featured two dinosaurs with boxing gloves facing each other. She didn't like the confrontational nature of the illustration, and the gloves were later removed at her request. The event attracted a huge crowd. The *Bradenton Herald* reported that there were more than a thousand people. Madrigal doesn't know exactly how many showed up, but she recalls that "there were buses and huge numbers of people coming in."[9]

By the time the debate was half over it was clear that the audience favored Gish. He got the majority of the applause, and spectators were blunt when asked by a *Bradenton Herald* reporter how Madrigal did. "I think they should have gotten someone a little bit more qualified," one said about her. "I felt good about my talk and about the way I communicated," Madrigal recalled, "but as the evening proceeded I realized that the event was not an educational experience but a rally for these people. I did feel used and ridiculed."[10]

Newspaper accounts of the debate highlighted a few of Gish's main talking points: evolution is just as much a religion as scientific creationism is, there is no evidence of transitional forms in the fossil record, and evolution was just as possible as a tornado blowing through a junkyard and creating a whole airplane. According to news accounts, Madrigal concentrated on how the fossil record supports evolution and on scientific discoveries about speciation. However, she now says that there was quite a bit more to the debate than was reported. For instance, during an audience questions segment, someone asked why there are no human fossils in the same rock layers as dinosaur fossils. Madrigal recalls that Gish had no answer for that. She also remembers Gish claiming that the probability of life arising was incredibly small, but Madrigal countered with: "We know what the probability of it happening is: It is one, because it did happen."[11]

Overall, Madrigal felt that Gish and the event organizers were respectful toward her, and she remembered that Gish was pleasant and polite. The crowd was also generally polite toward her during the debate, but once she was offstage during a break, the reception was icy. "I had to go to the restroom and was stared down by all the very unfriendly women in the line," she said. Madrigal looks back on the experience with some regret. "I achieved nothing by participating in it," she said. "I did not educate a single person. Never again. I wish I had not done it."[12]

A month later, though. Madrigal was back in the spotlight when she discovered that local public school teachers had been given continuing education credit for attending a creationism seminar. The eight-hour seminar was hosted by the Institute for Creation Research and was held in conjunction with the debate. When she found out that both the Manatee County and Sarasota County school systems offered their teachers the credit, Madrigal wrote letters to the school boards. She felt that it was her duty as a scientist to let the school boards know that creationism isn't science.[13]

Her letter was signed by eleven other scientists, including the director of anthropology graduate programs at the University of South Florida, Alvin Wolfe. He took offense to the inappropriate use of Madrigal's participation in the debate, which was considered part of the seminar. "Using one of our faculty in a program that offers credit without our faculty being told—and for an event that was set up to advocate religion—is wrong," he said. The scientists' letter stated that granting credit for the creationism event encouraged teachers to include the subject in their lessons. However, officials from both school districts saw nothing wrong with awarding teachers credit for attending. "Teachers need to see both sides of the argument in order to respond to questions from students," said one.[14]

John Morris, representing the Institute for Creation Research, said that the letter demonstrated "insecurity, arrogance and censorship." He said that the seminar was strictly scientific and wasn't religious. "Evolutionary theory is the keystone of the politically correct doctrine," he said. "And I think that is one of the primal evils in our country."[15]

"Exploring All Issues in a Balanced Way"

In 1993 the issue of creationism popped up in Osceola County when Marla Hughes, a concerned mother of three, asked the school board to consider including it in the classroom. The fundamentalist Christian wanted it taught alongside evolution. "I say put all the guesses on the table and let the kids decide for themselves which ones they believe," she said.[16]

She initially got encouraging feedback. School board member Joe Shirah had no problem with having creationism in the curriculum as long as it was deemed legal. Larry Brown, the school board attorney, felt that it might be permissible as long as it was taught "as a theory" and not as a religion. Board chairman Michael Harford added his support as well. "I'm for exploring all issues in a balanced way," he said. On the other hand, Superintendent Chris Colombo was skeptical. "Every time it's been done anywhere else, it's caused problems," he said.[17] Board member Donna Hart was also cautious, wanting to know if anyone beside Hughes was advocating this.

Before the subject was officially presented at a May 25, 1993, school board meeting, Assistant Superintendent for Curriculum Roger Dearing polled teachers at two middle schools and two high schools to see how evolution was taught. The district didn't have a strict policy in place concerning the subject, allowing teachers to do what they wanted. However, the informal poll revealed that no middle school teachers covered evolution or creationism. The majority of high school teachers also avoided both subjects, with only a few briefly mentioning them. "Most of the science teachers don't get into it because of controversy," Dearing said.[18]

Hughes was joined at the school board meeting by local doctor John Rao and activist Kathy Wolf. Rao had written a book titled *Think! Man! Think! Is God Real? Is Evolution True?* Wolf was known locally for her defense of a cross displayed on a city water tower. The end result of the board meeting discussion was that Hughes would submit a list of textbooks for review. However, any approved books would merely be resources for teachers, not required material. "Our job is to teach science in school," Dearing said. "So far we have not found a science of creationism."[19]

"It's Not Like It's a Whacked-Out Theory"

Despite staying in the public eye through public debates and school board meetings, Florida creationists weren't making much headway in the political arena. In 1994, state representative Buddy Johnson, a republican from Plant City, introduced House Resolution 2755: "A resolution recognizing creationism instruction in schools as a choice." Johnson said that he wasn't trying to mandate the teaching of creationism, but just wanted to give teachers permission to discuss it in class along with evolution. "It resolves it is not illegal to teach scientific creationism by virtue of a Supreme Court ruling, but it doesn't say we have to," he said.[20] Johnson wasn't able to round up any support, and the resolution was one of a handful that didn't pass.

The school board election season kicked off in 1994 with races in four counties featuring support for creationism. In Pinellas County the watchdog group Americans United for Separation of Church and State included in a candidate survey the question "Do you believe that creation science should be taught in science classes as a theory equal to evolution?"[21] Candidates Warren Andrews and Joe Triolo answered yes. They both said that evolution and creationism were theories on the same footing and so deserved to be taught together.

Andrews, a retired engineer and substitute teacher, assured voters that he wouldn't actually attempt to insert creationism in the curriculum, but he didn't see why the subject was excluded. "There's an awful lot of emotion in this," he said. "But even those people who are rabid for evolution still call it the theory of evolution. . . . Scientifically neither can be proved. It takes just as much faith to believe in one as in the other." Triolo claimed that creationism was a theory that had been around for centuries. "It's not like it's a whacked-out theory," he said.[22] Both men lost their elections.

Candidate Pattie McBrady battled in a Republican primary runoff in Lake County, where just three years earlier a prominent fight to insert creationism into the classrooms had failed. Unlike the candidates in Pinellas County, McBrady took a bold stance on behalf of creationism. "Offer creationism first and use Darwin as the alternative," she said. "Let them know where the Earth came from."[23] She was defeated in the runoff.

In Palm Beach County, David Walsh advocated for creationism in his school board race. "Why not present it to our kids and let them make up their own minds?" he asked. "Whether it's valid or invalid, it's the other side of the coin. A lot of people believe in creationism." Walsh targeted his opponent, Diane Heinz, as a person who told different audiences what they wanted to hear. Heinz was fine with creationism during a debate at a Christian radio station, he said, but at another forum she stated something different. Heinz countered Walsh's accusations by saying, "Creationism should not be taught as a science. Is that clear enough?"[24] Voters chose Heinz.

In Hardee County, political newcomer Patty Murray was interested in seeing if there was a way to teach creationism in the schools along with evolution. "So many textbooks that are provided today use evolution as if it were a fact instead of a theory," she said. "If they're going to teach a theory like evolution, then at least offer the other side of the coin."[25] She lost.

Of course, these candidates' campaigns didn't hinge on creationism; it was merely one issue among many that were addressed. However, it was typically a part of a grander conservative platform that was endorsed by local chapters of activist groups like the Christian Coalition. For instance, Cornerstone was a Florida organization that advocated "traditional values" and had sent questionnaires to candidates asking if they supported "creation-science as one alternative theory to evolution theory."[26]

Finding the Middle Ground

"The doors have been opened for the Body of Christ to minister on the school campuses," Jim Way wrote in a letter to St. Lucie County evangelical pastors in December 1994. Way was executive director of the Business, Civic and Ministry Coalition of St. Lucie County, and he and several religious leaders had sat down with Superintendent David Mosrie a few times over the previous months. The coalition hoped to convince Mosrie to allow creationism to be taught alongside evolution in the classroom. By the time January 1995 rolled around, Way really liked what he had heard from Mosrie. He claimed that Mosrie was fine with allowing teachers to present creationism if they wanted. "It's a theory that's got scientific validation, that there was some planned purpose to the way creation came about," Way said.[27] Since there appeared to be a green light from Mosrie, the coalition bought copies of the book *Of Pandas and People* to distribute to the local schools.

But a big problem cropped up in January. Reporters hadn't known about the situation until they were invited to a meeting between Mosrie and another group, the Port St. Lucie Ministerial Association. But that group's members weren't in complete agreement on the issue and weren't comfortable with the media's presence, so the reporters were asked to leave. But after several months of avoiding scrutiny, the issue was finally public. The school board members claimed they had no idea what Mosrie had been up to until they read about it in the newspaper.[28]

The Business, Civic and Ministry Coalition had met with Mosrie about half a dozen times since the beginning of the school year with the goal of giving creationism equal time with evolution. Way was confident that the model of creationism he wanted in the classroom had scientific validity and was not religious. "It's important to teach creationism in science classes because it's fact," he said. Another coalition member, Martin Drummond, a local minister, thought that omitting creationism was unfair and unprofessional. "I think if they're not giving a level playing field to creationism that gives the impression to students that creationism is bunk," he said.[29]

Way was confident that his efforts were moving along smoothly based on Mosrie's feedback. According to Way, Mosrie had already let teachers know that teaching creationism was permissible. The message he got from Mosrie was that he was "open to looking at, and understanding, all scientific data." However, Mosrie said that Way had possibly misunderstood. He had sent a letter to teachers about the issue in September, but he said that it only asked them to give students with various religious backgrounds some leeway when evolution was brought up. Mosrie, who had started his education career as a high school science teacher in 1969, said his letter merely instructed teachers to ensure that evolution was presented as a theory.[30]

The *Fort Pierce Tribune* obtained a copy of the letter, which read, in part: "Be sure that the theory of evolution is taught as a scientific 'theory,' not a law or fact. In that regard, I would encourage teachers to invite discussion of various points of view, if students desire." In the newspaper's view, this statement implied that alternatives to evolution, such as creationism, were allowed to be included in lessons. Mosrie said that regardless of how that statement was interpreted, the bottom line was that the risk of the school district being sued was too high, so now he was advising instructors to not teach creationism.[31]

However, Mosrie further muddied the waters by saying that legal threats could come from either side. The American Civil Liberties Union of Florida stated they were willing to sue if creationism was incorporated into lessons, but Mosrie claimed that creationism supporters could sue, too, "for not allowing equal time." Avoiding a lawsuit didn't mean that creationists' views should be dismissed outright, though. "I think with enough calm discussion, we can find that middle ground," he said. Other mixed signals may have come from Mosrie's agreement to let the coalition submit *Of Pandas and People* for review for use as a reference source in the schools. Mosrie hadn't read the book at the time, but had no problem with information about its general subject matter being available to students. "If this is handled properly, different viewpoints should not be a threat to anybody," he said.[32]

Once word about the meetings finally went public, the coalition found that they had a supporter for their cause on the school board. Board member Tom Coss said that covering only evolution was unfair. Instead, it should be balanced with equal time given to creationism. "I don't have a problem with students knowing there are two theories," he said. None of the other board members revealed their stances on creationism. All of the board members complained about having been kept in the dark, though. "The board was caught off guard," said Chairwoman Karen Knapp. "We didn't know we had gone this far with the ministerial association."[33]

The board members were also surprised to learn that copies of the book *Of Pandas and People* had already been sent to the principals and some teachers of all the high schools and one middle school for their review. Feedback from the teachers was unfavorable, with some saying the book was unscientific and "does not belong in science class." One high school biology teacher was blunt in her assessment: "It's really a book on creationism that's pretending that it's not." The board members' first action was to recall all of the *Of Pandas and People* books so they could review the text themselves.[34]

Poor communication seemed to be a chronic problem. At one point Mosrie said that the school board's involvement wasn't needed yet. "If there is a policy level decision that needs to be made, that's when it needs to go to the board," he said. But the next day the spokesperson for St. Lucie County schools, Chevon Baccus, said that all board members actually had been told in October via phone about the book distribution. It was further revealed that there had been no documentation kept about any of Mosrie's meetings with the coalition. Mosrie defended this conduct, saying that he saw them as just informal discussions with concerned constituents. As more news stories appeared and the controversy caught fire in the public, Mosrie pointed out that such a reaction was exactly why he had kept the issue private. He said that it wasn't a public policy matter and that rational discussion on the issue just wasn't possible due to its "great sensitivity."[35]

When reporters asked if teachers were already teaching creationism, the school district produced conflicting statements. "They can if they want," Baccus said. "I don't know. I expect many of them probably have." She explained further, saying, "Are teachers going to tell students that God created the earth? No, just as they don't tell them that people evolved; they tell them that's one of the theories." This ran counter to a statement by Queen Townsend, assistant superintendent for curriculum, that creationism wasn't part of the official science curriculum in the school district.[36]

"If They Want to Sue Me, They Can"

Nearly a week after the news first broke, the issue was finally aired out during the January 24 board meeting. Creationism wasn't on the agenda, but it nonetheless dominated the meeting. First, Mosrie corrected misinformation he said was being spread by the media. He said that his position on the matter had been distorted and that he certainly had not circumvented the school board. According to meeting minutes: "Dr. Mosrie emphatically stated that he did not and had never intended to require creationism be taught as part of the secondary science curriculum." However, he went on to ask the board members to not censor teachers if they permitted creationism discussions with students who might have varying points of view. Teachers shouldn't be in the position of telling students who believe the biblical creation story that they are wrong. Mosrie believed that teachers were professionals who should be trusted to handle the issue in an acceptable manner.[37]

The chairman asked school board attorney Dan Harrell to report on his research and offer advice. Harrell explained that there was no motion before the board to actually act upon, so his brief would be just for informational purposes. Case law was clear that creationism could not be included in the science curriculum, and *Of Pandas and People* wasn't an acceptable textbook. Harrell vouched for Mosrie's claim that there never was any intention to require creationism instruction. On the other hand, teaching and discussing the subject were two different things. If it came up in class, there should be no reason to avoid it provided the teacher was sensitive to students' various views.[38]

Coss was still the only school board member to take a stance. None of the others offered an opinion during the meeting. Coss said that he saw nothing wrong with students being told there are origin theories other than evolution. "I feel students should know that there are two theories. It's America, and if they want to sue me, they can."[39]

Way was blunt and direct when he spoke during public-comment time. According to the meeting minutes, he asked the board to "restore openness and academic honesty in our schools." He accused the ACLU and the media of intimidating Mosrie. However, his message was countered by the next speaker, Jim Mory, a local Protestant pastor. "The question of science is verifiability," he said. "That same word cannot be used with creation science. The word is a misnomer. Creationism is not science."[40]

Although the meeting ended without any specific action being proposed, the issue was kept alive in the media for a few more days. A story in the Sunday edition of the *Fort Pierce Tribune* told of the tensions that teachers sometimes deal with when it's time to cover evolution. High school biology teacher Mary Gregory explained that more than a decade prior she had been a teacher in Massachusetts, where, she claimed, there was a law requiring equal time for creationism and evolution. During her time teaching in St. Lucie County she was careful not to make her students too uncomfortable. She acknowledged to them that creationism does exist but that it wasn't a scientific theory. Sometimes, though, her students let their strong views be known. "I have had kids bring in books on creationism, telling me I better get my act together," she said. Other students told her: "My parents say I don't have to even listen to you talk about evolution."[41] She recognized that it could be an emotional subject for the students, so she patiently listened, didn't debate the issue with them, and kept on teaching.

In the same Sunday paper, a guest column by Mosrie was published. In it he explained that he had been approached by people who "felt ignored by our public school system." He said that he listened to them and then helped them find ways to "correct this antagonistic feeling" through the same procedures everyone else uses. The fact that this particular group consisted of ministers wasn't a reason to ignore their concerns.[42]

At a Christian Coalition of St. Lucie County meeting the next month, members were encouraged to send notes of thanks to Mosrie for his stance on allowing creationism to at least be discussed in classrooms. Public-policy chairman Bob Vogel didn't want to stop there, though. He said that his organization would still try to get equal time for creationism. However, later in the year the enthusiasm for the effort faded. In October, Way and several other board members of the Business, Civic and Ministry Coalition resigned. Way had been the driving force behind the five-year-old organization's projects, and without him the coalition faltered.[43]

"Biological Change over Time"

At the height of the creationism furor in St. Lucie County, the state's new education commissioner, Frank Brogan, also stuck a toe in the choppy waters. He was a strong proponent of local school district control and had no problem with St. Lucie or any other county permitting the presentation of creationism as an "alternate theory." He was aware that "evolution is a widely accepted theory," but he also said, "I think there is another belief out there and it can be taught."[44]

As Brogan settled into his job he never made a big deal of the issue. An opportunity to do so came, though, when something he did make a big deal over came to fruition: state education standards. The brand-new Sunshine State Standards, approved in 1996, established a clear set of academic expectations that were the same for all students no matter where they were in the state. But in keeping with Brogan's philosophy, school districts had local control over how they would achieve those standards.

Having statewide standards was a new concept in Florida, so it attracted a lot of attention, especially since all students would eventually have to pass mandatory exams based on those standards. Since there was so much focus on this dramatic change in state education, something that could have been controversial slipped by largely unnoticed. The new science standards didn't use the word "evolution" at all. In its place was the phrase "biological change over time." According to Tom

Baird, a Florida Department of Education policy analyst at the time, this was done on purpose. There was a fear that using the word "evolution" would ignite a cultural battle and sap precious time and energy that couldn't be spared. Baird said that anyone looking at the replacement phrase would know what it meant. They could then avoid "the outcry from church groups and so forth," he said.[45]

"Evolution Is Science Fiction"

Even though there hadn't been much success in getting it into the public schools, creationism was still a booming business as the 1990s gave way to the 2000s. Florida creationists worked hard to reach kids outside of the school campuses through churches, homeschooling families, and their own museums.

For instance, Kent Hovind, a former high school science teacher turned creation-science evangelist, had established his Creation Science Evangelism ministry in 1989. He produced several creationist videos and publications and traveled the world giving creationism seminars.

Creation Museum building at Dinosaur Adventure Land, Pensacola, 2008. (Courtesy Ebyabe, Creative Commons license via Wikimedia.)

He billed himself as "Dr. Dino" and claimed to have advanced degrees, including a Ph.D. in education. However, research into his academic credentials cast doubt on the validity of his degrees and revealed that he has no actual scientific background or worthwhile training in education. His seminars centered on answering questions like "Does science contradict the Bible?," "Did you know that dinosaurs are mentioned in the Bible?," and "Is there a political reason evolution is being promoted in the public schools?"[46]

In 2001 Hovind built Dinosaur Adventure Land in the backyard of his Pensacola home. Its dinosaur theme was a marketing ploy meant to attract the target audience of children and their families. The small park and museum had a science education veneer, but they were actually vehicles for Christian proselytizing, promotion of "young Earth" creationist beliefs, and teaching the idea that humans and dinosaurs had lived at the same time. "There are a lot of creationists that are really smart and debate the intellectuals, but the kids are bored after five minutes," Hovind told the *New York Times*. "You're missing 98 percent of the population if you only go the intellectual route."[47]

The park got Hovind into trouble, though, when he claimed he didn't owe the government any taxes or permit fees. Hovind said that the property and all the employees served God and thus weren't under any governmental authority. He went so far as to always pay the workers in cash. The IRS finally raided Hovind's properties in 2004, an act that he waived off as the government needing to "flex its muscle." Two years later Hovind was arrested for not paying $473,818 in taxes. He claimed he was being persecuted for his religious beliefs, but a jury saw it differently. They convicted him on fifty-eight counts of tax fraud, and Hovind was given a ten-year prison sentence.[48]

Dinosaur Adventure Land closed in 2009, but the following year Hovind's son opened a creation-themed store at a new location. In addition to selling "apologetics-based resources," the Creation Store offers tours and events. "Come to The Creation Store where you and your friends will learn how dinosaurs and the Bible meet."[49]

Hovind wasn't the only member of Creation Science Evangelism to spread the organization's message, though. In September 2000, Mike Schultz ran a seminar in Charlotte County where he gave public presentations several nights in a row while also speaking at more than a dozen area schools, both public and private, during the days. He flaunted Creation Science Evangelism's $250,000 prize that would be paid to anyone who could prove evolution was true. "Science deals with reality," he said. "What you can see, touch, taste, smell. Evolution is science fiction. It's fantasy pushed as scientific fact."[50]

The Creation Studies Institute is another Florida-based business with an active outreach program. A significant part of its mission is to develop "a strategy to combat the dogmatic teaching of evolution in public schools by exposing the brainwashing, censorship, and deception that is taking place." It has a Creation Discovery Museum in Ft. Lauderdale that houses fossils and other artifacts for public viewing. The institute was founded in 1988 by Tom DeRosa, who had taught science at a Ft. Lauderdale Christian private school and taught chemistry at Broward Community College. He's known for his Ice Age Fossil Adventure trips on the Peace River near Arcadia, where participants are "taught how to collect and interpret Florida fossils using a biblical framework."[51]

DeRosa also takes his large fossil collection on the road. He has frequently made presentations in Broward County public schools, but without any references to his creationist beliefs. On the other hand, DeRosa gives creationism seminars at churches and Christian private schools, and he makes a special effort to get his educational materials in the hands of homeschooling parents. "You're not just teaching kids, you're teaching moms and dads, too," he said during an Orlando home-schooling convention in 1998. "They're very anxious to learn."[52]

Another creationist organization that goes fossil hunting on the Peace River is Arcadia-based Creation Adventures Museum. Dr. Gary Parker and his wife host regular week-long Creation Education Vacations. A participant invitation letter explains: "We do His work presenting History from His Word in the Bible and in the world with the visual tools of science." Parker also gives presentations to Christian private school groups and homeschooling families. Among the things he has taught is that *Tyrannosaurus rex* was an herbivore rather than a carnivore. Its massive teeth cracked coconuts instead of bones.[53]

Parker had worked for several years for the creationist organizations Answers in Genesis (based in Kentucky) and the Institute for Creation Research. He also taught biology closer to home at Clearwater Christian College. Before then, he says, he was an "evolutionist" with a biology/

chemistry degree from Wabash College, Indiana, and an education doctorate in biology from Ball State University. According to Parker, his scientific knowledge and research led him to start doubting aspects of the theory of evolution.[54] He has written several books, including *Life before Birth: A Christian Family Book* and *What Is Creation Science?* coauthored with Dr. Henry M. Morris from the Institute for Creation Research.

In 1991, Lake County had a very public tussle over a request to teach creationism. Hoping to help convince the school board to allow the subject, Grady McMurtry offered his experience as an evolution-believing atheist turned creation-believing Christian. In 1994 he founded Creation Worldview Ministries, headquartered in Orlando. He has since traveled around the world spreading his creationist message, but he doesn't neglect his home state. McMurtry provides a long list of presentations for church groups to choose from, most of which feature some aspect of creationism. He firmly believes public schools should teach creationism: "When they teach only evolution, it is indoctrination, not education."[55]

Homegrown creationists weren't the only ones making a lot of noise in Florida around this time. In 1998, Answers in Genesis held a major event in Vero Beach, the first time the group held a large seminar in Florida. Local residents who formed a committee to bring the seminar to their town hoped it would excite and motivate people to get more active in demanding creationism in the schools. In the promotional materials, the message was clear: "Let's reach the students and teachers in our schools with the truth, and deprogram them from evolutionary brainwashing!"[56]

8

"There Are Razor Blades in That Apple"

It was 1996, and a few concerned Lee County residents felt something was lacking in their public schools. They believed students needed exposure to traditional values and character education, and a great way to get that was through elective Bible classes. The National Council on Bible Curriculum in Public Schools, based in North Carolina, offered a way to address residents' concerns. The organization said that their curriculum taught about the Bible from historical and literary perspectives without crossing the line into proselytizing. Three hundred schools in eight states were using the curriculum, according to the company, but none were in Florida. Lee County could possibly be the first.[1]

Beverly Kehn, who said she was a local volunteer for the National Council, informally proposed the idea to the Lee County School Board in January. The board members seemed receptive, so Kehn then presented it to the board's curriculum advisory committee. Once again, she found the majority of the members there liked it. There was some concern that time and money shouldn't be spent on new electives when some students still struggled with basics like reading and writing. Others felt that the Bible was strictly a religious book. "It's not literature, it's the Bible," said one committee member.[2] Nonetheless, on a 6–4 vote, the committee approved the basic concept of having two courses, Old

Testament and New Testament.[3] That cleared the way for the school board to formally discuss the courses during a workshop, followed by a possible vote at a regular meeting in late March.

The U.S. Supreme Court had ruled in 1963 that schools couldn't use the Bible to teach religion, but they could use it in a strictly secular manner. The National Council on Bible Curriculum assured the Lee County School Board that its curriculum had been vetted by lawyers and was entirely in keeping with the Supreme Court's ruling. But there was a fine line to walk, and people on both sides of the issue worried that things could go wrong. "It all depends on who teaches it," said Janice Quinlan, a member of the Concerned Women for America, a Christian women's watchdog group. "You get a hold of an evolutionist, hmph. And believe me, there are some who would try to do it. You know, they can be very biased." On the other hand, curriculum committee member Terri Wampler couldn't see any way to separate the Bible from religion if the National Council's curriculum was used. "I see Genesis, Adam and Eve, creation," she said. "How do you teach that without religion? I don't know how you can teach this without mentioning God."[4]

School board meetings were typically sparsely attended, but the late-March meeting was packed with at least 250 citizens wanting to debate the issue. After nearly seven hours of public comment and board members' discussion, a 4–1 vote permitted the creation of the two Bible classes. Three weeks later, the school board, superintendent, and district curriculum director started the nomination process to appoint seventeen members to a citizen committee that would piece together the actual curriculum to be used. The final committee featured businessmen, retirees, parents, a minister, and a rabbi. Even though the National Council offered their curriculum for use, the board wanted their own locally created one.[5]

As 1996 came to a close, the contentious issue took a serious toll in Lee County. Crafting a curriculum was slow and difficult work that involved a lot of debate and compromise. For instance, the committee members needed to decide which Bible translation students would use. After quite a bit of discussion, they finally settled on using a book that contained four translations. In October, the committee reported to the school board that the Old Testament curriculum was nearly complete. But in December the curriculum committee got mired in a stalemate because a contingent insisted that the content of the Old Testament class would prompt lawsuits. Then school board member Doug Santini saw another possible reason for the curriculum difficulties. He presented to his fellow board members a list of complaints about Superintendent Bobbie D'Alessandro's performance, including his belief that she was dragging her feet in implementing the Bible curriculum. The board eventually agreed to buy out her contract, essentially firing her.[6]

Another official to fall victim to the conflict was school board attorney Steve Butler. He had counseled the curriculum committee that the Bible program developed so far "plainly contravenes the Constitution." School board member Bill Gross sent him a letter afterward, scolding him that his inappropriate comments "may have upset that delicate balance which the committee has worked to achieve." Butler promptly resigned in January 1997. But shortly after his departure, two law firms that specialized in First Amendment religious rights cases nationally issued statements that despite their support of Bible classes in general, Lee County's ran a serious risk of being unconstitutional.[7]

Just two weeks after their own lawyer resigned, the school board hired a St. Louis, Missouri, law firm at $145 an hour to review the course materials to keep the school district out of legal hot water.[8] However, in April the law firm expressed concerns about the legality of certain learning objectives from the Old Testament course. The curriculum committee made revisions, but not enough to avoid continued criticisms.

"Young People . . . Don't Know Who Our Creator Is"

The Old Testament course revealed in the summer of 1997 had a clear fundamentalist Christian flavor. The story of Adam and Eve was presented as "universal history," and students would be expected to "list the days of Creation," "find out what was created on each day," and figure out how many animals of varying sizes Noah's ark could hold. This sparked fears among the class's opponents that the conservative majority on the school board was trying to "establish a creationist academic curriculum."[9] However, the course had yet to be finalized, largely because of fierce disagreements among committee members, intense pressure from opposing factions in the community, and multiple threats of legal action.

In contrast to the passionate community interest, the students for whom the courses were intended had little interest. Of roughly thirteen thousand high school students who could sign up for the electives, only about two hundred did so.[10] Meanwhile, the national media—including National Public Radio, the *Washington Post*, and CNN—were paying close attention to what happened in Lee County.

On July 15 1997, just a little more than a month before the Old Testament course was to be taught for the first time, a heavily revised version went before the school board for review. "The attorneys have sucked the marrow from the Bible," said resident Connie Holzinger. "Miracles are now 'actions called miracles,' sin cannot be discussed, and Adam and Eve are banished from the course, just as they were banished from the garden." Committee member Bill Bracken was also unhappy with Adam and Eve's expulsion. Creation is "historic and not just religious," he said. "If you leave them out, what history would you have? If there was no God, you would have no history." But Keith Martin, Butler's replacement as the school board attorney, said that those objectives had to go. Including the lessons "would be to teach the creationist belief," he said. "And that would subject us to the argument that we are teaching beliefs, not history."[11]

At the August 6 meeting, the school board faced an electrified crowd of more than two hundred people. Over the course of two and a half hours, seventy-five speakers expressed their opinions, with about two-thirds of them in favor of the Bible curriculum. School board member Lanny Moore, who had recently won his seat on the board with the support of the local Christian Coalition, made it clear why he was behind the Bible course: "I would submit many young people in our (schools) don't know who our creator is." Upon hearing that, one resident felt sure that Moore had just stirred up unneeded trouble by showing that he had a religious motive. "I think he put himself and the board in a very precarious state regarding any future litigation," she said. But on a 3–2 vote, the board approved the Old Testament course, giving it a green light for the upcoming school year.[12]

The controversy was ratcheted up yet another notch when principals in the only two schools that had planned to offer the new course canceled it. There wasn't enough time to properly train the teachers, they said, especially since they would face so much scrutiny and pressure, and not all of the course materials and texts had been selected yet. The new plan was to offer the class in the second semester. However, advocates of the Bible class accused the principals of having an agenda. "This is a denial of freedom of choice, and it is a censorship of views," Kehn said. Nonetheless, no Bible courses would be offered in any Lee County schools until January 1998. In the meantime, the teachers tapped to lead the Bible classes finally got thirty-one hours of training. This included instruction on legal issues as well as tips on how to work with the media.[13]

"Created in God's Image or Created from Slime"

But before the chronically delayed courses could finally kick off, the inevitable happened. On December 9, 1997, People for the American Way and the American Civil Liberties Union of Florida, working with Florida law firm Steel Hector and Davis, sued the Lee County School Board. They claimed that the curricula were unconstitutional and filed a preliminary injunction to prevent any such classes from being taught in January. The religious liberties group American Center for Law and Justice offered to defend the school district pro bono.[14] With prominent and well-funded organizations ready for battle on both sides of the issue, it was predicted that the impending federal court case would even gain prominence in the national news.

On Tuesday, January 20, 1998, Judge Elizabeth Kovachevich of the U.S. District Court for the Middle District of Florida ruled on the injunction request. The Old Testament course could proceed because of the careful curriculum editing that lawyers had imposed. However, the New Testament course, which had not been subjected to the same scrutiny, was temporarily banned. The teaching of the Old Testament course would need supervision, though, by means of videotaping. The seven high schools offering the course were supposed to start the classes Wednesday, but this new requirement forced a one-day delay so that recording equipment could be obtained and set up. Then another difficulty arose. Lawyers had hoped to see the videos the day after each lesson, but school principals said the tapes could not immediately be released due to student privacy concerns.[15] When the tapes were finally released, attorneys would review them and present a status report within sixty days to Judge Kovachevich.

Merely a week after Old Testament classes began, a new storm appeared on the horizon. During a regular meeting of the curriculum advisory committee, members discussed the benign subject of ensuring that the county's core curriculum was aligned with the state's Sunshine State Standards. However, when they reached the high school science sections on evolution and the Big Bang Theory, some members suggested that other theories should be taught, too. "If we are going to be intellectually honest with our children, we should let them look at all theories," said committee member Bill Bracken, who had also been a member of the Bible curriculum committee. "I don't understand why they would not want to let (students) know. Isn't that what education is all about?"[16] He and other members of the committee believed there was nothing wrong with allowing students to compare creationism and evolution.

Support was likely available on the school board for this newest effort. Board member Moore hadn't left anyone guessing as to his stance during a campaign forum that McGregor Baptist Church hosted in 1996. "If you believe you descended from the apes or slime, that affects you," he said. "For the first 150 years, creation was taught and we didn't have the (social) problems that we have today. It matters if you believe man is created in God's image or created from slime."[17]

Teaching creationism in biology classes never got past the idea stage, though. Lee County's curriculum executive director, Doug Whittaker, was certain the issue wouldn't gain any traction. "In other places where they've tried to do it, it has met with much opposition," he said.[18] The Bible courses were soaking up all of the energy and attention, so the suggested changes to biology classes never came to fruition.

Finally, in February 1998, a settlement was reached concerning the Bible courses. The school district agreed to dump the curricula the board had approved and replace them with a purely secular set of courses based on the college-level textbook *An Introduction to the Bible*.[19] The book was written by professors at Stetson University in DeLand, Florida. The Bible would be examined from a literary point of view, with no claims that its content was factually true. Furthermore, those who filed the lawsuit reserved the right to monitor the classes and take the school district back to court if there were any agreement violations.

The fallout eventually reached the individual board members during the next election cycle. The majority that had pushed the Bible issue was reduced to one person as the voters took the board overall in a decidedly moderate direction.[20] Lee County's time in the media spotlight was finally over.

"The Truth of Creation"

It had been seven years since creationism last popped up in Manatee County. But history repeated itself there in 1999 when Rev. Gary Byram took up the cause. His first appearance before the school board was on November 1, when he asked to have creation science taught in the schools. He was supported by board member Frank Brunner. According to the meeting minutes, Brunner was eager to get to work on the issue: "He requested that a Workshop be scheduled with the Curriculum Department, for experts in this area to present a true science curriculum that gives equal time to evolution and creationism in the classroom."[21]

Part of what emboldened the men to support an issue that had been defeated in Florida and in their own county so many times in the past was the success other states were having in 1999. In Kansas, new testing standards minimized evolution's importance, and the word "evolution" was replaced by "changes over time" in the state standards. In Oklahoma, disclaimers in biology textbooks, saying evolution was a "controversial theory," were approved by a state textbook committee.[22]

Byram had started Bible Baptist Church in the late 1970s in Bradenton. The fifty-six-year-old was inspired to stand up for creationism when famous creationist speaker Kent Hovind had visited his church in 1993. Byram invited Hovind back in November 1999 to headline three seminars and timed his request to the school board to coincide with the event. "We'd like to expose people to the truth of creation," Byram told the *Bradenton Herald*. "It's taught biblically, but we believe it's also scientifically accurate and should be taught in the schools."[23]

Byram went before the board again on November 16, the day before his seminars were to kick off. The subject of creationism had yet to be included on the official board agenda, so the board members didn't discuss it. However, the district's curriculum staff was researching creationism per Brunner's direction at the previous board meeting, according to the district's assistant superintendent of academics, Lynette Edwards. "We are reviewing all the issues surrounding the teaching of evolution and creationism," she said. "We are not taking any position at this time."[24]

Hovind attracted a crowd of 650 people for one of the seminars at the Bradenton Municipal Auditorium, including Brunner.[25] In fact, Brunner had worked with Byram to bring Hovind to Manatee County. He helped promote the event and raise money to cover some of its expenses. This upset some citizens, but his actions didn't cross any legal or ethical lines. Brunner was participating as a private citizen, not as a school board member, and creationism had yet to appear on a board meeting agenda. However, the National School Boards Association cautioned him to consider recusing himself should the issue ever come before him on the board.[26]

But Brunner then stirred up even more controversy. He directed that a school board employee tape Hovind's talk and make copies available to the curriculum staff for review. Brunner defended this, telling the *Bradenton Herald*: "I don't think it's a religious one (issue). I think it's a matter of science." Brunner also had no qualms about giving his opinion on the matter. "Personally, I would like our school system to teach creationism side-by-side with evolution or any other theory that's out there that's supported by scientific evidence," he said.[27]

Joining Brunner in favoring creationism was fellow board member Barbara Harvey. "I feel that it should be taught as an alternative theory," she said. "I'm a Christian, and I know creationism is right." The other three board members were noncommittal, but that didn't dampen Byram's enthusiasm. He showed the school board some creation science books he hoped to include in classrooms, and he planned to meet one-on-one with Brunner to get things moving. "This is just getting rolling right now," Byram said in late November. "But I think there's a great empathy that can be generated in Manatee County."[28]

Creationism still hadn't made it onto the agenda in December, but Brunner and Byram continued their efforts. Brunner realized that there were legal issues that would need to be worked out. "I would like to see somebody bring me a curriculum that has withstood legal challenges in another district," he said. Apparently, that wish never came true. Byram spoke at two more school board meetings, but he failed to get anything done. Board members echoed Brunner's concerns about legality, saying that the lingering threat of litigation would definitely stop the process before it even started. Even if there was a legal way to get creationism into the classroom, board member Larry Simmons didn't want to get involved. "It's an issue that can drive a major wedge in this community and, at this time, we don't need that."[29]

The issue eventually fizzled out. The next time it came up was during the county's 2000 school board elections. Harry Kinnan had been one of the school board members who had mostly kept quiet during the brief creationism push. Although he didn't come right out and say that creationism should be taught in the classroom, he appeared to be leaning that way. "I think any reasonable theory should be exposed to children," he said. His challenger, Clint Chapman, was more forthright. He said that science "absolutely" could be used to justify creationism.[30] Kinnan won.

"They Want a Christian Education"

The importance of education to Floridians was highlighted in 1998 when citizens voted to strengthen the education clause in the state's constitution. Previously, the clause simply said, "Adequate provision shall be made by law for a uniform system of free public schools and for the establishment, maintenance and operation of institutions of higher learning and other public education programs that the needs of the people may require."[31] The new version states:

> The education of children is a fundamental value of the people of the State of Florida. It is, therefore, a paramount duty of the state to make adequate provision for the education of all children residing within its borders. Adequate provision shall be made by law for a uniform, efficient, safe, secure and high quality system of free public schools that allows students to obtain a high quality education and for the establishment, maintenance, and operation of institutions of higher learning and other public education programs that the needs of the people may require.[32]

The change was intended to "make the Legislature finally do its duty by the public schools or be forced to do so by the courts," according to a *St. Petersburg Times* editorial column. Also in 1998, Jeb Bush won the gubernatorial election, and soon after being sworn in to office he followed through on a campaign promise to reform Florida's public education system. His "A-plus Plan for Education" emphasized standardized testing. Each school's students' test results would be compiled and calculated along with other school performance measures to produce an overall letter grade. That grade would then mark individual schools in a very public manner as passing or failing. State funding of schools was also tied to the grades.[33]

In 1999, Governor Bush pushed the legislature to approve a controversial component of his "A-plus" plan: the Opportunity Scholarship Program. This was a voucher program that allowed students who attended consistently failing public schools to either transfer to a higher-performing public school or use state funds to attend a participating private school. Opponents of the program claimed that "a right-wing crusade is afoot to undermine public schools and redirect state funds to private, religious-based schools."[34] Bush signed the "A-plus" education bill, which included the scholarship program, into law on June 21 of that year. Florida was the first in the nation to implement a statewide voucher program.

The Opportunity Scholarship Program wasn't the only voucher effort under way in 1999. The McKay Scholarships for Students with Disabilities Program was also part of the "A-plus" education bill. But the McKay Scholarship didn't attract as much attention, because it started as just a pilot program in Sarasota County that year. In this program, vouchers were based on students' special needs instead of a school's performance. Eligible students could transfer between public schools or choose to attend a private school.[35] It didn't stay in one county for long, though. Bush signed a bill in May 2001 that expanded the McKay Scholarship program statewide.

Also in May 2001, a third voucher program, the Florida Tax Credit Scholarship Program, was implemented by the state to "encourage private, voluntary contributions from corporate donors to non-profit scholarship funding organizations (SFOs) that award scholarships to children from low-income families."[36] Businesses get state income tax credits for donating money to fund approved private school scholarships for qualified students.

In all three voucher programs, the participating private schools could be sectarian or nonsectarian. However, for a private school to be eligible to participate in the Opportunity Scholarship Program it had to meet a short list of requirements and "Agree not to compel any student attending the private school on an opportunity scholarship to profess a specific ideological belief, to pray, or to worship." However, this wasn't on the list of private school requirements in the other two voucher programs. In fact, the majority of the schools accepting them were religious. For instance, of all the private schools accepting McKay Scholarship students in the 2011–12 school year, 64 percent were religious and 36 percent were non-religious. In the same school year, private schools accepting the Florida Tax Credit Scholarship students were 77 percent religious and 23 percent non-religious.[37]

"Many of the parents bring their kids here because they want a Christian education," the principal of a voucher-accepting private school told the *Palm Beach Post* in 2005. "And a Christian education does not include evolution." Noting that the state did not track what curricula were used at private schools, reporters investigated. Their survey found that 43 percent of religious schools that accept voucher students used curricula from conservative Christian publishers. According to the *Post*, this amounted to "about 375 voucher-taking schools, educating about 8,700 students" statewide.[38]

One of those companies is A Beka Book, which is based in Pensacola and affiliated with Pensacola Christian College. Its science textbooks are based on biblical literalist and "young Earth" creationist beliefs. The *Post* pointed out that an eighth grade textbook sold by the company contains a chapter on "science versus the false philosophy of evolution." A sixth grade science textbook was advertised as follows in 2012: "This teachable, readable, and memorable book presents the universe as the direct creation of God and refutes the man-made idea of evolution."[39]

The day after the Opportunity Scholarship Program was signed into law by Bush in June 1999, opponents, including the ACLU and the National Education Association, sued to have the program stopped. They claimed that the voucher program violated both the federal and the state constitutions in multiple ways. One complaint was that providing voucher money to private religious schools violated Article I, Section 3 of the Florida Constitution, which states: "No revenue of the state or any political subdivision or agency thereof shall ever be taken from the public treasury directly or indirectly in aid of any church, sect, or religious denomination or in aid of any sectarian institution."[40]

The case, *Holmes v. Bush*, bounced back and forth among the Florida courts until November 2004, when an eight-judge majority of the whole Florida Court of Appeals determined that the Opportunity Scholarship Program did violate that state constitutional provision. It was also determined that the program violated another, separate provision that requires the state to provide a "uniform, high quality education." On appeal, the case went to the Florida Supreme Court, where it was determined in January 2006 that the vouchers were unconstitutional under the "uniform" education provision. However, the justices declined to offer an opinion on the state constitution's prohibition against providing aid to "any sectarian institution."[41]

The end result, though, was that Opportunity Scholarship vouchers could no longer be used at private schools. The other two voucher programs—McKay and Corporate Tax Credit—were not affected. As of this writing there has yet to be any legal challenges to them.

"Sedulously Avoided"

In March 1998, Florida was taken to task in a national report that assessed the value of several states' public school science standards. The Thomas B. Fordham Foundation, a private group based in Washington, D.C., that tries to combat what it calls a "dysfunctional or ineffective" U.S. public school system through research and projects, had reviewed and graded states' standards in core academic subjects. When the foundation got to the science standards, it found that Florida's were "seriously flawed." The report stated: "Sad to say, irrelevant or trivial examples and poorly or erroneously stated ideas are common in this document." This resulted in Florida's F grade along with eight other states. All science subjects were examined, including evolution's place in the life sciences. Florida took a serious hit here: "The word 'evolution' is carefully avoided. The issue is skirted and such matters as genetic variation and natural selection are treated lightly; biological evolution is certainly not treated as the central principle of the life sciences."[42]

The Fordham Foundation followed up with another report in 2000. *Good Science, Bad Science: Teaching Evolution in the States* narrowed the focus to just that one subject. "Twelve states fail so thoroughly to teach evolution as to render their standards totally useless," the report said.[43] Florida was on that list, earning the state another F grade.

Five years later the foundation reviewed all of the states' science standards again, this time widening the scope to encompass all science subjects. Since Florida's standards were the same as they had been in 1998 and 2000, they failed again. "The E-word is sedulously avoided," the reviewer noted. "There is little in the way of useful guidance for teachers or others toward appropriate content in the biological sciences and especially in the history of life and the basic mechanisms of change."[44] But the 2005 report also noted that Florida was due to rewrite the science standards. The original set, written in 1996, was scheduled to be reviewed by the Florida Department of Education in 2006. As state officials in 2005 eyed that approaching task, there were early signs of a brewing conflict over the teaching of evolution.

In the early 1990s, the teaching of intelligent design had been advocated in Florida through failed attempts to get the book *Of Pandas and People* into science classrooms in Manatee and Pinellas Counties. Although the teaching of intelligent design didn't gain much traction in Florida, it was receiving a lot of attention elsewhere. In 2005 a Pennsylvania school district was neck deep in a prominent court case involving the mandated mention of intelligent design in biology classrooms. *Time* magazine published a cover story on intelligent design. President George W. Bush publicly supported teaching intelligent design, saying "both sides ought to be properly taught." However, Florida governor Jeb Bush, the president's brother, avoided the subject as best as he could. He was able to dodge the question for a couple months, but in early October 2005 he was finally cornered and asked if intelligent design should be in classrooms. "I don't know . . . I don't know," he replied. "It's not part of our standards. Nor is creationism. Nor is Darwinism or evolution either."[45]

A few days later, the governor was told that he was wrong. The concept of evolution was in the state standards, but just not mentioned specifically by name. Bush claimed that he was fed the erroneous information by state education commissioner John Winn. Bush's next statement didn't do much to clarify his stance: "I like what we have right now. And I don't think there needs to be any changes. I don't think we need to restrict discussion, but it doesn't need to be required, either."[46]

Meanwhile, the hiring in August 2005 of a new chancellor of K–12 education, whose responsibilities include implementing standards for students and teachers, stirred up some controversy related to evolution and the science standards. Cheri Yecke had previously been Minnesota's commissioner of education in 2003, but she was forced out of the job when that state's legislature refused to confirm her after she had been on the job for a little more than a year. Her supporters in Minnesota praised Yecke for not settling for the status quo. She had passionately advocated for high-stakes testing and school choice while criticizing teachers unions and the system's inability to close the achievement gap. Yecke's critics, on the other hand, said she was polarizing and tried to impose her conservative views on the education system. When Minnesota revised the state's social studies standards, Yecke was accused of trying to give them a markedly right-leaning and highly patriotic flavor. She didn't want a "hate America" curriculum in the schools.[47]

When Minnesota's science standards were being revised in 2003, Yecke had come under fire again. Wesley Elsberry, a Florida native and information project director for the National Center for Science Education, had kept track of the controversial process and said that Yecke persisted in inserting language in the standards that downplayed evolution. He said that she also tried to use a failed amendment to the federal No Child Left Behind Act that referenced alternative theories to evolution as a tool to influence the standards. "The only thing she was interested in was the intelligent design issue," recalled one teacher who had worked with Yecke on the revisions. "The other 95 percent of what we do in science was ignored."[48]

After only three months on her new Florida job, Yecke found herself defending her Minnesota job performance to the media, denying all of the negative claims. She said that she didn't allow her personal views to influence her professional life. Creationism was a "non-issue" in Minnesota, and she would not allow it to derail Florida's science standards revision process either. But it became an issue in 2007 when Yecke hired an Internet company called reputationdefender to find incorrect information about her and have it corrected or removed. At the time she wanted to move up to the position of Florida education commissioner, but she claimed that hiring reputationdefender had nothing to do with her career aspirations.[49]

One issue Yecke asked the Internet company to look into was a 2003 newspaper article that contained information she disputed. The article reported that Yecke issued a statement when she was Minnesota's commissioner of education saying that "schools could include the concept of 'intelligent design' in teaching how the world came to be." Elsberry wrote about the quote on his website, giving it further public attention. However, Yecke said she never made that statement, and she had reputationdefender contact Elsberry to ask that he either change or delete the reference. He refused. Instead, he told the Florida media about Yecke's hiring of the Internet company, which resulted in Yecke's reputation-cleaning attempts backfiring by drawing yet more attention to the issue. "Florida is primed for the sort of large-scale evolution/creation incident that has grabbed headlines in other parts of the country," warned Elsberry.[50]

"The Classroom Dictator"

"That's a healthy time to have discussions of that nature," said state representative Dennis Baxley when asked about intelligent design and the science standards revision.[51] He was chairman of the House Education Council, and in early 2005 he fired a warning shot of sorts before the process of rewriting the standards even got started.

Baxley was no stranger to controversy. He had been at the forefront of the Florida legislature's effort to prevent Terri Schiavo's feeding tube from being removed in the emotional 2005 right-to-die debate that attracted intense national attention. During that same legislative session, Baxley, a funeral director and son of a Southern Baptist minister, introduced House Bill 837, The Academic Freedom Bill of Rights. It was his attempt to protect college students with conservative views from being discriminated against by liberal professors. Baxley said that his bill would prevent "biased indoctrination" by "the classroom dictator" by giving students the ability to demand classroom discussion of

Rep. Dennis Baxley (R-Ocala) gestures while debating in favor of a bill on the House floor on March 17, 2005. The bill, later approved by the House, was an attempt by the legislature to intervene in the Terri Schiavo right-to-die legal case by giving the governor the authority to reinstate Schiavo's life support. (Mark Foley, State Archives of Florida.)

their views without being punished via bad grades or other academic retaliation.[52]

The bill didn't actually originate with Baxley. He modeled it on a version written by activist David Horowitz, founder of Students for Academic Freedom, a conservative think tank with the goal of combating the predominance of leftists at universities. Similar versions of the bill were also introduced in other state legislatures. What made Baxley's version unique was the personal story that motivated him to take up this cause. When he was a student at Florida State University in 1970, an anthropology professor went on a "tirade" against creationism. "I just slumped down in my seat and kept quiet, spit back what they wanted, ducked my head and got out of there," he recalled.[53]

With that episode still fresh in his mind, Baxley wanted to empower students to stand up to a "dogmatic professor" who might say: "I don't even want to hear anything about creation or intelligent design. And if you don't like any of that, there's the door." Baxley found himself having to backpedal a bit, though, and make assurances that he wasn't promoting the teaching of creationism. Instead, he just wanted to stop the belittling of conservative students' views in college classrooms. "We never let it happen to an African-American," Baxley said. "We never let it happen to some other ethnic group. We'd never let it happen to an Islamist. But we have no problem if you want to let loose on a conservative Christian Republican born-again."[54]

The bill got an icy reception from the state's universities as their presidents pointed out that procedures were already in place to handle reports of discrimination. Tom Auxter, president of the United Faculty of Florida, saw the dark side of the bill: "It presents itself as a shiny red apple that defends academic freedom, but there are razor blades in that apple." He also had sharp words concerning Baxley's anthropology class memories. "If he had a bad experience 30 years ago, get over it," he said. There was some confusion concerning what the bill would allow students to do if they were faced with a "dictator" professor. One official House of Representatives bill analysis document said that students would be able to sue, while a separate analysis said that the bill didn't allow for litigation. The bill was also filed in the Senate, where an analysis said that the bill would allow lawsuits.[55]

The bill easily passed through a few House committees, but then it stalled on the House floor, where it eventually died. It never even got a hearing in any Senate committees. As it became clear that the bill was failing, Baxley took advantage of all the attention it had garnered, along with the power of his position on important House education committees, to have a meeting with the state university presidents. To prove his point about the ridicule conservatives endured, he showed the presidents a cartoon that was published in the University of Florida student newspaper. It depicted the classic "ascent of man" evolutionary chain with a naked Baxley sitting behind a monkey. Although the presidents were polite, they appeared unmoved. "I don't think you have this bastion of liberalism that people think," said one.[56]

"It's So Innocuous"

As the academic freedom bill sputtered out, attention shifted to another fireworks show: public school textbooks. It had been six years since the state had last purchased new science textbooks, and just like in the case of the upcoming standards revision, there was apprehension that evolution's coverage in the books would cause problems. Committees across the state were created to review books submitted by textbook publishers eager for lucrative contracts in the big state. Florida's choices could also influence what materials would be offered to other states later. Once the committees whittled the potential list down, Commissioner Winn made the final decisions in November. School districts could then choose from the state's approved list of funded materials or venture out on their own but have to pay for the books themselves.

On the state level the approval process ran smoothly. It was when the school districts had to choose books that the ride got bumpy. A list of state-selected books was offered for the districts to choose from, but one of those books became a hot potato due to two paragraphs. *Biology: The Dynamics of Life*, published by Glencoe, included the following passage:

> Common to human cultures throughout history is the belief that life on earth did not arise spontaneously. Many of the world's major religions teach that life was created on Earth by a supreme being. The followers of these religions believe that life could only have arisen through the direct action of divine force.
>
> A variation of this belief is that organisms are too complex to have developed only by evolution. Instead, some people believe that the complex structures and process of life could not have formed without some guiding intelligence.[57]

Additionally, the teachers edition suggested letting students debate life's origins, including both scientific and religious explanations. Some school districts were turned off. Palm Beach County's science curriculum supervisor, Fred Barch, put it bluntly: "We talk about things we can measure, things we can observe. We don't get into other explanations that involve religion."[58]

But other school districts didn't see what the problem was. Broward County's science curriculum supervisor, J. P. Keener, felt that acknowledging religious beliefs could lead to worthwhile discussions. "Some teachers are so reluctant to talk about it," Keener said. "I don't know why." Broward County School Board member Stephanie Kraft liked the two paragraphs and was even hopeful that the board would consider allowing teachers to cover intelligent design. She felt that religious views on origins were just as valid as scientific ones. "Personally, I don't understand how evolution works," she said. "I don't understand how you went from one cell and then all of a sudden you got man." Her enthusiasm was countered by fellow board member Bob Parks, who supported evolution-only instruction, saying the book controversy appeared to be "another part of someone's political agenda." Glencoe responded to the hoopla in a message stating that it had not intended to promote any particular religious view. The passage was merely there to provide teachers with material for an optional critical thinking exercise.[59]

In early December, Broward County's High School Science Adoption Committee—composed of parents, teachers, and district staff—narrowed the choices down to the Glencoe book and *Biology*, published by Holt, Rinehart and Winston. It was now up to the district's teachers to decide which one they wanted in the classrooms for the next six years. Superintendent Frank Till then announced that Glencoe had offered to remove the controversial page if his district chose their book. "They agree with us that in the whole context of that textbook, it doesn't fit," Till said. "And it was probably a mistake to put it in there." Despite that concession, the district's teachers chose Holt's *Biology*. News coverage noted that several other school districts also passed over Glencoe's textbook, such as Pinellas, Hillsborough, Duval, and Palm Beach.[60]

The offer to take the paragraphs out was also extended to Brevard County. The district's textbook committee unanimously recommended to the school board to accept, but board member Amy Kneessy thought otherwise. "It's so innocuous," she said. "To me, those two paragraphs belong there." She said that she wasn't advocating for creationism in the classroom, but she thought everyone was being hypersensitive about it. "We're making the subject taboo, and I think that's wrong."[61] When the board was set to approve all of the textbooks at their March 14, 2006, meeting, Kneessy requested that they discuss the two versions of *Biology: The Dynamics of Life*. She explained that all of the ideas on the disputed page were valid and that students were merely being asked to examine them and form their own opinions. The meeting's minutes showed she was alone, though:

> Bea Fowler stated she is concerned that discussion will be held in science class led by science teachers and not by comparative

religion teachers. This subject should be taught by trained comparative religious teachers and not science teachers. Larry Hughes agreed that the place for this subject is in a comparative philosophy or comparative religion class. Janice Kershaw stated she would support the committee's recommendation. Robert Jordan stated he also believes the statement is worrisome and he would support the recommendation.[62]

Kneessy lost 4–1.[63] The committee's and superintendent's recommendation to use the version without the passage carried the day.

"Give Them All the Information"

A related issue had simultaneously arisen in Marion County. The district had been discussing intelligent design for about two years and decided to buy DVDs and books on the subject for placement in the libraries of all seven of its high schools. The book, *Of Pandas and People*, and the DVD, *Unlocking the Mystery of Life*, were not to be used as actual classroom instructional materials. They were meant as resources for students completing assignments or doing their own research. "We're dealing with it as a controversial issue," Superintendent Jim Yancey told the *Ocala Star-Banner*. "We're just trying to be open and honest and address what it is."[64]

The state's science standards seemed to support Yancey. Commissioner Winn had written a memo in October 2005 explaining that the standards "are neither inclusive nor exclusive to any one theory of human origin. However, the standards do encourage students to seek information for themselves by researching and exploring a variety of theories." Winn had written the memo in response to repeated questions about how evolution would be handled in the upcoming state science standards revision. "For me or anyone else within the Department of Education to state a personal belief on this subject, or to presume what may or may not be included in the revised science standards, would be nothing short of negligent," he wrote. "I categorically refuse to succumb to irresponsible media requests to prematurely opine on a topic before it is appropriate."[65]

Nonetheless, it was clear that organizations and individuals across the state were getting ready for a big fight over what should be in the science standards. On one side were those who felt there was room for intelligent design. For example, most members of the Collier County School Board felt that it would be fine to at least mention it in class. "We want our students to be well-rounded," said board member Steven Donovan. "We should give them all the information." "I guess you could say I'm a creationist," said a national board-certified science teacher in Polk County. "I always tell the students human beings are awesome to me. There has to be something that designed all this." "There's so much detail down to the cell level," a Pasco County biology teacher said. "In my opinion, it would be impossible for this to all have been a coincidence."[66]

On the other side were those who felt that evolution should be in the science standards without intelligent design. "Evolution is a theory supported by facts that can be proved. You don't have that with intelligent design," said a science supervisor in Alachua County. "Evolution is not a dirty word," said a University of Central Florida professor who prepared science teachers. "I don't think you can be a true biologist without believing in evolution." A November 2005 Florida Association of Science Teachers Conference included a workshop titled "Darwin under Fire: Classroom Strategies for Teaching Evolution in a Climate of Controversy." The Florida Association of Science Teachers had also adopted the National Science Teachers Association's position statement on evolution, which stated that creationism and intelligent design were discredited.[67]

In November 2005, a *St. Petersburg Times* poll of local parents found that 60 percent had paid some or a lot of attention to the controversy over evolution and intelligent design. Of those who had knowledge about the issue, 58 percent felt that intelligent design should be taught just like evolution, while only 21 percent didn't want intelligent design taught at all.[68]

But just as the issue really built up some steam, the science standards revision was suddenly postponed. In December 2005 the department of education announced that the revision process, which was due to start in 2006, would be delayed by at least a year. The reason given was that ongoing updates to the math and language arts standards had taken longer than expected, pushing back the revision of the science standards. There was some speculation, as noted in the *St. Petersburg*

Times, that the real reason for the delay was the fact that gubernatorial and state legislature elections were coming up and that the evolution debate was just too explosive.[69]

Although Governor Bush would be leaving office, the media hounded him for a clarification of his stance on teaching evolution. In late December 2005 he attempted an answer. "I am a practicing Catholic and my own personal belief is God created man and all life on earth," he said. "However, I do not believe an individual's personal beliefs should be the basis for determining Florida's Sunshine State Standards." But later in his statement, Bush said: "Perhaps more importantly, we should encourage the vigorous discussion of varying viewpoints in our classrooms. A healthy debate of issues challenges our students' minds." This encouraged Mathew Staver, president of the Orlando-based conservative legal organization Liberty Counsel, who told the *Orlando Sentinel*: "I think what Jeb Bush is stating, as I read his statement, is that he is open to having a robust debate on the issue of evolution take place in the classroom. That's all intelligent-design advocates are asking for."[70]

Others thought the governor's stance had been anything but clarified. "Really, there's no reason to be vague," said state representative Dan Gelber, a Miami Beach Democrat. "Do you think it should be in science classes or not?"[71] But that was the last time Bush directly publicly addressed the issue before leaving office. Chancellor Yecke also moved on without getting entangled in the evolution controversy, resigning in December 2007 after failing in her attempt to become the state education commissioner. There was some question, though, as to what influence Bush still had over the process, since he had selected all but one of the state's seven board of education members.

Preparing for Battle

With evolution education in schools a hot topic yet again in Florida, the citizen group Florida Citizens for Science was formed in late 2005. The organization was modeled after similar groups in other states and was closely affiliated with the National Center for Science Education in California. The group's main focus was on networking across the state to build support in preparation for what members felt would be a major clash when the science standards were revised. Members also successfully secured various positions on the state committees that would plan and write the standards.

Also hoping to influence the upcoming standards revision process, a new Clearwater-based group called Physicians and Surgeons for Scientific Integrity (PSSI) was formed in May 2006. The group advocated for the teaching of intelligent design and promoted a petition called "Dissent from Darwinism" signed by doctors "skeptical of nature-driven Darwinian macroevolution."[72] The petition was modeled after a one that the Discovery Institute had used in national campaigns in an effort to discredit evolution.

In late September, PSSI hosted a highly publicized intelligent-design conference called "Darwin or Design?" at the University of South Florida's Sun Dome sports arena in Tampa. It featured a trio of leading national authorities on intelligent design: Michael Behe, Ralph Seelke, and Jonathan Wells. High school and university students and university faculty got free admission, and students were given a free copy of the DVD *Unlocking the Mystery of Life*. Attendance estimates varied from one to four thousand. The University of South Florida's student newspaper, *The Oracle*, observed: "The crowd was largely composed of non-USF students wearing clothing reflecting their Christian beliefs."[73]

The event's local organizer was Dr. Thomas Woodward, a professor at Trinity College in Tampa Bay. He had completed his doctoral work in the Department of Communication at the University of South Florida, where he wrote his thesis on the history of the intelligent-design movement. That work later became the book *Doubts about Darwin: A History of Intelligent Design*. Woodward also founded the C. S. Lewis Society.[74]

9

"I Want God to Be Part of This"

A committee of thirty-three "framers" met in May 2007 to finally kick off the official standards review process. The Office of Mathematics and Science—a branch of the Florida Department of Education—assembled science educators from all levels along with business leaders and private citizens with the purpose of helping to decide what should be in the new document. The framers heard from nationally recognized experts and examined national and international research. They then created guidelines for the next group of twenty-five "writers" to use in creating the first draft of the new science standards. In October, the writers turned in their product.[1]

The standards draft was significantly changed from the 1996 version. For instance, the subject matter was divided up and presented as "big ideas," such as "Forces and Changes in Motion" and "Earth in Space and Time." These ideas could be explored in depth rather than the old standards' method of presenting a wide range of scientific concepts that could only be given surface-level treatment in the limited time available in the school year. One highlight was that evolution was a "big idea." Dr. Lawrence S. Lerner, who evaluated state science standards for the Thomas B. Fordham Foundation and had given Florida's standards an F in previous years, called the draft "a giant step in the right direction. . . . It is clear, comprehensive, and most importantly, accurate."[2]

Only one person raised any significant opposition to evolution's place in the standards draft during their development. Fred Cutting, a retired aerospace engineer in Clearwater and a member of the framing committee, stated his concerns in a letter he called a "Minority Report." Although he considered the new standards draft a "great improvement," he complained that the proposed standards took a "dogmatic tone" that did not "reflect the true nature of science, and dramatically overstate the degree of proof supporting Neo-Darwinian evolution and theories of chemical evolution." He recommended a handful of edits, but the framing committee didn't use them.[3]

Even though Cutting's protest didn't get results, it was significant that he had enlisted the aid of the Discovery Institute, the primary promoter of intelligent design nationwide. Institute member Casey Luskin wrote on the organization's blog: "Some time ago, Mr. Cutting inquired with us for information about solid evolution education, and we were happy to supply it, along with input on his draft Minority Report."[4]

It was relatively smooth sailing for the committee during the six-month draft-creation process. The next step was to invite public input. The draft standards for other academic subjects had received minimal public interest, but this time the response was overwhelming. The Office of Mathematics and Science posted the draft science standards on a website and invited the public to rate and comment on them for sixty days. When the comment period ended in mid-December, the website had logged 262,524 ratings. In contrast, the math standards had elicited only about 43,000 ratings. Citizens could also leave anonymous comments on the website. Most comments that addressed evolution were favorable, such as: "It is very promising that you intend to introduce this concept at such an early age." But there were plenty that took an opposing view: "I am so sick that people have become so brainwashed into thinking that evolution is true."[5]

Evolution Dogma

As if the website wasn't keeping the Office of Mathematics and Science busy enough, four town-hall meetings were scheduled across the state to collect more citizen input. The first ones, in November, were lightly attended, and two were even canceled. One meeting at Jones High School in Orlando attracted forty people, but only ten spoke about evolution. Florida Citizens for Science president Joe Wolf was among the four speakers who supported the new standards. "Teaching intelligent design, creationism, can only cause confusion in the minds of students," he said. "How can we expect students to learn science when we're teaching religion?" At another November meeting, in Tallahassee, Wakulla County School Board member Greg Thomas said that the new standards were too "radical" and suggested that they be abandoned in favor of the 1996 standards. "This will run afoul of many students and teachers," he said.[6]

But then citizen interest picked up steam, as Florida Board of Education chairman T. Willard Fair discovered in early December. He told the *St. Augustine Record* that evolution in the standards had prompted more correspondence to him than any other subject. Fellow board member Dr. Akshay Desai also found himself buried under more than one hundred e-mails about evolution in just a couple of weeks. "I haven't seen this level of passion" on other education issues, he said.[7] The majority of the correspondence came from people advocating for intelligent design.

"Passionate" certainly described one of the most prominent anti-evolution voices to arise during this time. Kim Kendall was a "self-described mom-who-won't-go-away" from St. Johns County. She got her start in activism soliciting $50,000 worth of book donations from businesses for her children's elementary school. The former air-traffic controller moved on to become a prominent voice fighting her local school district's rezoning efforts and protesting new home construction in an area where the roads were already congested and dangerous. She helped launch the American Eagle and Literacy Challenge, which was a program to raise money for American bald eagle protection while also promoting literacy. "When I do something, I do it 200 percent," she said.[8]

Fresh off these campaigns, Kendall wanted to make sure her opinions about evolution's place in the science standards were heard. She accused the department of education of trying to get the science standards approved as quietly as possible in order to avoid any controversy over evolution. Kendall said that the public's input on the science standards was being ignored and that the public hearings held so far were poorly advertised, while others she had planned on attending were canceled. She wanted to talk directly with the seven state board of education members, since they were the ones who would be making the final decision. State senator Stephen Wise tried to help her by securing for Kendall a time to address the board at its regular meeting on December 11, even though discussion of the standards was not on the meeting agenda.[9]

Despite the senator's intervention, Kendall was told the night before the meeting that she wouldn't be given time to speak. The state's education commissioner, Eric Smith, who was about to preside over his first board meeting since taking over for Commissioner Winn, had contacted her directly. He explained that the denial was based on fairness to all and that there would be more appropriate times available later. Kendall and a few others attended the meeting anyway, since they had driven two hundred miles to be there. According to the *Florida Baptist Witness*, their trip wasn't a complete waste, as they still were able to meet with individual board members "to share their concerns."[10]

The department of education's original plan was for the board to consider the standards during a meeting in January. But then it was decided to schedule two new public hearings that month, so the board wouldn't vote on the issue until its February 19 meeting. One hearing was scheduled for January 3 in Jacksonville, and the other was five days later in Miramar.

The decision makers Kendall desperately wanted to talk with had managed to stay out of the spotlight for a couple of months, but in early December one finally made her views clearly known. State board member Donna Callaway exchanged e-mails with the executive editor of the *Florida Baptist Witness*, James A. Smith Sr., in which she expressed her dissatisfaction with the new standards. In an editorial, Smith broke the news: "Florida's educational establishment is attempting to make evolution dogma the sole means of understanding the origins and development of biological life for students in the Sunshine State, and Florida Baptists—and other concerned citizens—should be participating in this debate." His article went on to explain that Callaway had no problem with evolution being in the standards but that other theories about life's origins should also be included. "This has the possibility of confirming or denying for a child who he/she really is," Callaway had written to Smith. She didn't think that intelligent design should be taught, but she believed that teachers should be allowed to "acknowledge that there are other theories."[11]

"With All Its Warts"

Callaway was a mother of two who had been a language arts teacher and media specialist for several years. She moved up to an assistant principal position and finally served as a middle school principal for ten years in Leon County. Gov. Jeb Bush appointed her to the state board in 2004. It was noted in the *Witness* that she was a member of First Baptist Church in Tallahassee. She was quoted in Smith's piece as saying: "My hope is that there will be times of prayer throughout Christian homes and churches directed toward this issue. As a SBOE [state board of education] member, I want those prayers. I want God to be part of this. Isn't that ironic?" The *St. Petersburg Times* took Callaway to task in an editorial, saying that she should step down. "When a member of the state Board of Education puts her religion before the educational needs of Florida students," the *Times* wrote, "she forfeits her standing as an education expert and should resign her post."[12]

Another person who wanted Christian action—albeit more than prayer—was Selena Carraway, a program manager for the Florida Department of Education's Office of Instructional Materials. She sent an e-mail on her own time from her personal e-mail account explaining to recipients that schools would have no choice but to teach evolution if the new standards were approved. She wrote: "Whose agenda is this and will the Christians in Florida care enough to do something about it?" Of course, she was well within her rights to send the e-mail, but what got her in trouble was that she had explicitly detailed in the e-mail what her position was in the department of education in order "to give this e-mail credibility." When word of that got back to her bosses, she was reprimanded for advocating for a personal cause using her official position.[13]

With one state board member's opinion finally revealed, a few others also let the public know which side they were on. Linda Taylor went on the record as sympathetic to the teaching of alternative theories alongside evolution. "I would support teaching evolution, but with all its warts," she said. "I think that some of the facts have been questioned by evolutionists themselves. I would want them taught as theories. That's important." Roberto Martinez firmly planted his flag on the pro-evolution side: "I'm a very strong supporter of including evolution. And I think it's long overdue." Dr. Desai's public statement was more muddled. On the one hand, as a medical doctor, he believed in evolution. But he also wasn't completely opposed to giving creationism a mention in the standards. "There is a significant passion about this issue from a religious perspective," he said. "That needs to be respected."[14] The opinions of the other three board members—Fair, Kathleen Shanahan, and Phoebe Raulerson—remained unknown.

"I Didn't Come from an Amoeba or a Monkey"

The board of education wasn't alone in facing the brewing controversy. Local school districts also found themselves in the spotlight. The first to grab attention was Polk County. When school board member Kay Fields was asked by the local newspaper for her opinion on the science standards, she replied that she was going to consult with her superintendent about what could be done at the district level. "There needs to be intelligent design as well," she told the *Lakeland Ledger*. "You need to show both sides." A follow-up story reporting on a poll of the school board members found that five of the seven supported Fields's views. "If it ever comes to the board for a vote, I will vote against the teaching of evolution as part of the science curriculum," board member Margaret Lofton said. She added, "It crosses the line with people who are Christians. Evolution is offensive to a lot of people."[15]

Shortly after the *Ledger* published its findings, the board members were inundated with correspondence from an unusual source: the Church of the Flying Spaghetti Monster. Followers of the Internet-based satirical faith initiated an e-mail campaign demanding that the Spaghetti Monster's version of intelligent design via the touch of its "noodly appendage" be taught. The board members didn't think the campaign was funny. "They've made us the laughingstock of the world," Lofton said.[16] Soon afterward the board members conceded that the science standards were the state's problem and that nothing done at the local level would have an impact on the process.

Meanwhile, school boards in the northern reaches of the state were taking action. Taylor County superintendent Oscar Howard Jr. mentioned at one of the January public hearings that his county and several others were sending official resolutions to the state board of education encouraging them to either deemphasize evolution or allow alternatives to be taught. Howard, who had driven nine hours to attend the hearing, claimed that hundreds of parents were threatening to pull their kids out of public schools if the standards were accepted in their current form.[17]

Many of these county school boards tried not to make a public fuss over their resolutions, with some only being publicized after Florida Citizens for Science discovered them through research in small-town newspapers and the minutes of school board meetings. At least a dozen counties are known to have passed resolutions: Clay, Jackson, Baker, Hamilton, Holmes, St. Johns, Taylor, Madison, Lafayette, Nassau, Washington, and Bay. The majority of these counties were in the northern and panhandle areas of the state, regions well known for their cultural and political conservatism. "I'm a Christian. And I believe I was created by God, and that I didn't come from an amoeba or a monkey," Madison County School Board member Ken Hall said.[18]

Many of these resolutions were nearly identical. They chafed at evolution being referred to as established fact, such as this Baker County resolution that was approved unanimously:

> WHEREAS, the Florida Department of Education has drafted and is now proposing new Sunshine State Standards for Science, the Baker County School Board opposes the implementation of the new standards as currently presented.
>
> WHEREAS, the new Sunshine State Standards for Science no longer present evolution as theory but as "the fundamental concept underlying all of biology and is supported in multiple forms of scientific evidence," we are requesting that the State Board of Education direct the Florida Department of Education to revise the new Sunshine State Standards for Science so that evolution is not presented as fact.
>
> WHEREAS, the Baker County School Board recognizes the importance of providing a thorough and comprehensive Science education to all the students in Baker County and to all students in the

State of Florida, it recognizes as even more important the need to present these standards through a fair and balanced approach. NOW THEREFORE, BE IT RESOLVED by the Baker County School Board of Baker County, Macclenny, Florida, that the Board urges the State Board of Education to direct the Florida Department of Education to revise the new Sunshine State Standards for Science such that evolution is not presented as fact.[19]

Several other school districts had brushes with the controversy. Putnam County had a resolution on the school board meeting agenda but then removed it from consideration prior to the meeting. Highlands County considered a resolution but abandoned it when several citizens spoke against it at the board meeting. Monroe County bucked the trend by passing a resolution in favor of evolution in the standards. Volusia County School Board members told their local newspaper that they were in favor of having evolution in the standards even though they never formalized that stance during an official meeting.[20]

The majority of Pinellas County's school board members opposed evolution, but they didn't attempt to express that in a resolution. "I'd probably ideally like to keep it all out of the classroom," said Peggy O'Shea. "If it's going to create this much controversy, how important is it?" Janet Clark, who was in the minority on the issue, emphasized how American students were falling behind other countries in science education. "Let's start teaching the Bible as science and then see how our students compete against the rest of the world," she said. Dixie County also didn't draft a resolution, but board members were in agreement on the subject during a meeting discussion. "We just wanted to get it on the record that we're a Judeo-Christian community, and we believe in academic freedom," said Superintendent Dennis Bennett.[21]

"This Is Like the Middle Ages"

However, as Polk County had concluded, the school districts weren't going to be casting votes in February. That was entirely up to the state board of education members, who were facing increased pressure as time wore on. The national evangelical Christian organization Focus on the Family urged citizens to contact members of the state board. In response, Florida Citizens for Science initiated its "All I Want for Christmas Is a Good Science Education" campaign. Floridians were encouraged to send Christmas cards that included short notes in support of sound science and evolution to board members.[22]

The anti-evolution forces were veterans at mobilizing for various conservative causes. Terry Kemple was president of the Tampa Bay–based Community Issues Council, which describes itself as "Christian Citizenship in Action." He often stood side-by-side with Kendall, the activist mom, at events protesting evolution in the standards. Kemple and Kendall also worked hand-in-hand with the Florida Family Policy Council—the local affiliate of Focus on the Family—headed by John Stemberger. The *St. Petersburg Times* referred to Stemberger in 2007 as "the loudest voice in Florida's Christian right movement."[23] His Florida Family Policy Council boasted of having tens of thousands of e-mail addresses and thousands of active donors.

The *Florida Baptist Witness* and Florida Baptist State Convention tapped into an equally vast network of enthusiastic supporters. These activists worked hard to dominate news coverage and tried to influence the state board of education through unrelenting pressure and sheer force of numbers. The Florida Family Association, another vocal conservative organization, took credit for more than 13,800 e-mails sent to board of education members asking to have evolution taught as a theory and not as a fact.[24]

Even though the anti-evolution players had impressive networking capabilities and could stir up tremendous support from the general public, the pro-evolution side had powerful allies, too. Among the organizations that gave support through letters, resolutions, and advice were the National Academy of Sciences, the National Center for Science Education, the American Institute for Biological Sciences, the Florida Academy of Sciences, Americans United for Separation of Church and State, and Florida members of the Clergy Letter Project. The battle attracted the attention of the ACLU, which sent a letter to the state board in which it cautioned against an unconstitutional endorsement of religion.[25]

Additionally, the standards' writers and framers didn't just walk away when the draft was done. They continued to advocate on the standards' behalf, and forty of them signed a letter sent to the board encouraging adoption of their version of the standards without giving in to outside pressure. Gerry Meisels, director of the Coalition for Science Literacy at the University of South Florida and a member of the standards drafting committee, presented the letter at one of the public forums. The letter warned that Florida would be viewed as a "backward state" and drive away businesses that depended on an educated workforce if those against evolution got their way. "It's very counterproductive for our children, it's counterproductive for our country, it's counterproductive for our future," Meisels said. "This is like the Middle Ages."[26]

Nobel Prize winner Dr. Sir Harold Kroto, professor of chemistry at Florida State University, made sure he cleared time in his busy schedule to help, writing in a newspaper op-ed:

> It is disgracefully unethical for individuals who rail against the teaching of evolution to young people as a proven "fact" to accept, either for themselves or their families, the humanitarian benefits accruing from medical scientific research underpinned by the theory. Evolution is the backbone of biology. Many medical treatments including most drugs could not have been developed if previous generations of young biology and medical students had not been taught evolutionary concepts.[27]

Much of the support for the science standards was only loosely organized, though. Florida Citizens for Science became the focal point of the coordination effort, and through its activities it built a foundation of hundreds of individuals willing to help. The group spearheaded a petition effort that gathered more than seventeen hundred signatures both on paper and on the Internet, including many prominent scientists and present and past Florida university presidents.[28]

"Indoctrination Centers"

Those on the anti-evolution side countered the ACLU's legal warning with one of their own. Kemple promoted a legal memorandum co-written by attorney David C Gibbs III and curriculum consultant Francis Grubbs. Gibbs was already well known for representing Terri Schiavo's parents and siblings in the 2005 right-to-die case in Florida and for representing Florida creationist Kent "Dr. Dino" Hovind in a tax fraud case.[29]

In their memo, Gibbs and Grubbs cautioned the board of education that the draft science standards appeared to be hostile to religion. The memo claimed that evolution as presented in the standards wasn't science but rather a "philosophical faith-based belief system." They warned that this could have legal repercussions: "We are concerned about the scientific accuracy of the Florida standards and also about the potential some of these proposed terms might have for requiring only one particular belief system in Florida classrooms, which would be an unconstitutional violation of the Establishment Clause."[30]

Evolution reared up in regional politics, too. Bill Foster, a former St. Petersburg councilman with aspirations to higher office, sent a letter to his local school board warning against the evils of evolution. "Evolution gives our kids an excuse to believe in natural selection and survival of the fittest, which leads to a belief that they are superior over the weak," he wrote. He connected evolution to Hitler and the Columbine High School massacre. A couple of years later Foster's creationist beliefs were brought up during his campaign for St. Petersburg mayor.[31]

The two newly scheduled public meetings in January were much more vocal affairs than the previous forums. About 120 people attended the one in Jacksonville, where almost all of the roughly 50 speakers had something to say about evolution. Kemple was on hand, saying: "This isn't about whether or not our children should be taught that evolution is a fact. The issue goes to whether our schools are places of learning or indoctrination centers." Kendall was also there repeating an argument she used several times as the debate grew. "How many of us were taught that Pluto was a planet?" she asked, referring to science's uncertainty.[32] She believed that evolution could eventually be cast aside, too.

A few days later in Miramar, the next meeting featured about thirty-five speakers, almost all of whom debated about evolution. Support for teaching evolution was notably high there. For example, a high school student lamented that it would be difficult to get into "an upstanding university without evolution in my textbook." Backing up the eleventh grader was state representative Shelley Vana, who had experience teaching science. "I taught [advanced placement] biology," Vana said. "You can't even teach that and have your kids pass the test if you don't teach evolution." On the opposite side of the issue was Oscar Howard, the Taylor County superintendent. "I think they could be teaching a lie," he said. "There's not a place on me where they took the tail off."[33]

"They Know We Have New Standards"

The mountain of feedback on the standards was compiled, and the writing committee met January 9–11 in Tallahassee to hammer out a final version for presentation to the board of education. Several changes were made in response to the citizen input. One standard that attracted a lot of negative response read: "Recognize and describe that fossil evidence is consistent with the idea that human beings evolved from earlier species." About 52 percent of people who rated that item were either in the "strongly disagree" or "disagree" category. That prompted the revision: "Recognize that fossil evidence is consistent with the theory that living things evolved from earlier species."[34]

The revised document was released for public viewing on February 1 and was immediately attacked. Despite some adjustments to the overall document, the basic evolution standards had been left mostly intact. Editor James Smith wrote in an editorial in the *Florida Baptist Witness*: "Like the first draft, what is missing from the revised standards is any recognition that there is controversy about Darwinian evolution and that students should learn about that controversy. Whatever happened to academic freedom and exposing students to all sides of a debate? For the evolution-as-dogma crowd, there is only one side when it comes to Darwin." Despite four public hearings, the invitation for public comment on the Internet, and the flood of e-mails and letters to the board of education, anti-evolutionists wanted another opportunity to have their say. "It's a good thing that so many people are concerned about science," said Mary Jane Tappen, executive director of the Office of Mathematics and Science. "At least we have their interest and they know we have new standards."[35]

So, a fifth public forum was held on February 11 in Orlando. Despite the fact that the new draft covered every aspect of science education in the public schools, all of the more than seventy speakers focused on evolution. News reports estimated that at least forty-five speakers opposed the subject, including this sample described in the *St. Petersburg Times*:

"One man linked Charles Darwin to Adolf Hitler, Josef Stalin and Mao Tse-tung. Another said evolution sanctioned murder. Still another held up an orange and said that because of evolution, he now had irrefutable evidence that an orange was 'the first cousin to somebody's pet cat' and 'related to human beings.'"[36]

Tappen was there along with Commissioner Eric Smith and K–12 chancellor Frances Haithcock for the entire five-hour meeting, but evolution opponents were angry that the board of education members weren't there. "All I want to do is have my voice heard and before the seven people who will make the decision," Kendall said. "They need to see the whites of the eyes of the parents who are affected," said Stemberger.[37] The anti-evolution coalition vowed to lobby the board members until that happened. The relentless barrage of requests worked.

Initially, public input wasn't going to be allowed during the February 19 board meeting, but the week beforehand that changed. Bowing to the pressure, the board issued a news release on Thursday announcing that a limited number of people could speak for three minutes each. The release stated: "It is anticipated that testimony will be heard from a maximum of 10 speakers in support of the standards as written and 10 speakers in support of the standards with changes."[38] Speakers would have to arrive the morning of the Tuesday meeting and sign up for those twenty slots on a first-come, first-served basis. This sent both sides scrambling to decide who should speak, what they should say, and how they could get there early enough to secure spots on the speakers list.

Adding to the stress in the final stretch was a dramatic and surprising eleventh-hour proposed change to the science standards. Department of education officials were nervous that the board would never approve the standards due to so much opposition to evolution. With little time in which to work, they hastily consulted with three biologists who had worked on the standards committees. The question was if the word "theory" could be added to the evolution sections in order to appease the tidal wave of complaints and yet still be academically sound. It was eventually agreed that the phrase "the scientific theory of" would be fine and that inserting it into relevant sections throughout the standards, not just in relation to evolution, would be reasonable and fair.[39] For example, an original statement in the standards read: "Explain how evolution is supported by the fossil record." After the change it read:

"Explain how the scientific theory of evolution is supported by the fossil record."

This compromise document was rushed together a mere week before the board meeting and was officially revealed on the afternoon of Friday, February 15, right before a holiday weekend. The *St. Petersburg Times* later obtained from the department of education hundreds of e-mails that detailed the sequence of events and the reactions. There was a wide range of responses from liking the changes to thinking the changes weren't enough. Some of the teachers and scientists who had created the standards were upset. "I recognize that the public unrest with the standards we have written may be disconcerting, however, to create an alternate version simply to please the masses, does not make the final result world class," wrote elementary school science teacher Janet Acerra.[40]

"A Process That Is Fair"

Now the board members had the choice of approving the standards as originally written (Option A); approving the last-minute "scientific theory of" compromise version (Option B); or rejecting the new standards. Of course, the board could do any number of other things if they wanted. They could rewrite entire swaths of the standards at the meeting if the majority desired. But it was likely that the board would accept one of the three main options.

As the sun rose on February 19 in Tallahassee, opposing factions lined up in the state capitol building to get on the speakers list. Both sides were organized and prepared. Once the doors opened at 8:30, the room quickly filled. There were only about 120 seats, and once they were all taken, many people were left standing. Several reporters dominated an entire section of the room, and TV cameras were packed in like a platoon of soldiers.

The meeting started promptly, and the first few general items on the agenda were handled. Shortly after 9 a.m. the standards discussion began. Chairman Fair opened with a short speech that seemed to be aimed at the anti-evolution audience members. Sometimes he even spoke directly to Kendall, who was sitting in the front row. He made it clear that the standards public-review process had been carried out in a transparent manner with several properly advertised opportunities for everyone to have input. After getting more than two thousand e-mails in just five days, Fair joked, "I want to thank those persons who sent them to me because now I know that my Blackberry works."[41]

However, as Fair looked right at Kendall, he said that some people wanted to speak directly to the board and look them right in the eye. "We responded, Kim, to the notion that you needed to look at me and see my body language and understand whether or not I was listening to you as you made your presentation," Fair said. "We wanted to make sure that if we accommodated you it would be fair, and we think that we've set up a process that is fair."[42] Thus, the opportunity to speak at the board meeting was now offered.

As if foreshadowing the battle to come in the next few months in the state legislature, a trio of lawmakers was given time to address the board first. House Minority Leader Dan Gelber cautioned the board not to mix science and religion. "The truth of the matter is if you start to put faith and religion into science class you do neither of them any service," he said. He then advised them to reject the last-minute compromise version of the standards: "You've added 'scientific theory' all over the place. . . . You've changed them somewhat dramatically on a Friday night. And I don't know what they will do; whether it will be a wink and a nod to teachers to do something that they probably ought not be doing. But I do know that those new optional standards have not been vetted. They haven't gone through the process you spent a year setting up."[43]

Rep. Ed Homan said he was confused as to why the controversy had gotten to the point it had, since he didn't see a conflict between creation and evolution. A baby is a miracle of creation, but God made it happen using evolution. "Why are we arguing about teaching the scientific principles of the way our creator made this happen?" he asked. Next, Rep. Marti Coley advocated for Option B. Students needed to study both the "strengths and weaknesses of any information," she said. She stressed that nothing had been taken out of the standards and that they had not been watered down. "We are simply asking that you use the word 'theory' in conjunction with the word 'evolution' to acknowledge that there are many unanswered questions."[44]

The twenty citizens then had their chance to speak. Comments alternated between anti-evolution and pro-evolution sides. The pro-evolution speakers primarily tried to explain how objections to teaching evolution had a religious foundation. For instance, Florida Citizens for Science's Jonathan Smith emphasized that science itself is not important to those opposed to evolution. Their goal is to undermine it so as to support their particular religious ideology. As proof, Smith referred to the recent federal case in Pennsylvania in which intelligent design was found to be religious in nature and promoted by those with purely religious motivations. He also quoted from the writers of the "Wedge Document," a Discovery Institute goal-setting paper outlining how to "defeat scientific materialism." The primary strategy was to cast Darwinism as "inherently atheistic."[45]

Retired Baptist minister Rev. Harry Parrott referenced the Clergy Letter Project, a petition-type project where thousands of clergy members across the nation signed a statement agreeing that evolution does not conflict with their faith. Those in opposition to evolution, though, see no middle ground, he said. "It is religious believers who have been fussing about this and debating about this for decades and decades and it will continue," Parrott told the board. "Do not let this religious debate intrude into our science classes and do not let this religious debate drive your decision today."[46]

Gerry Meisels, a chemistry professor at the University of South Florida, was one of the most eloquent speakers for the standards, drawing on his experience as one of its authors. He explained that there had been a discussion about how to best address the important ideas of scientific inquiry. The decision was to write a "Nature of Science" section rather than have unnecessary duplication in every subject's section. However, the concepts in the "Nature of Science" section were not meant to be taught separately; instead, they were meant to be integrated into all other instruction throughout an entire science course. Meisels referenced employment studies and surveys which showed that too much of the state's workforce was not educated enough to fulfill minimum job qualifications. Downplaying science in favor of "other perspectives" would be a "disservice to our students" who will later have to be remediated in college, he said. Meisels asked the board to trust the experts who had created the standards.[47]

Dr. Harold Kroto, professor of chemistry at Florida State University and Nobel Prize winner, reminded the board that Florida's government had been investing heavily in getting biomedical research companies to set up shop in the state. Kroto was on the Scripps Research Institute's board of governors and emphasized: "Evolution theory is fundamental to what Scripps does."[48]

The other six pro-evolution speakers were lawyer T. R. Hainline; evolutionary biologist Joe Travis, dean of Arts and Sciences at Florida State University; pastor Brant Copeland of First Presbyterian Church in Tallahassee; orthopedic surgeon Ray Bellamy; Monroe County School Board member Debra Walker, an anthropologist and a member of the science standards framing committee; and Dr. Paul Cottle, professor of physics at Florida State University, who was also on the framing committee. They all had their own points to make but emphasized the theme of keeping religion out of science classes, too.

The anti-evolution speakers were against both the draft standards and Option B. Kendall was the first speaker for the anti-evolution side, and she introduced a concept new to the Florida evolution debate. She presented to the board a single-page document titled Academic Freedom Proposal. It asked that a simple change be made to the standards. The definition of "evolution" as presented in the draft science standards was:

> Evolution is the fundamental concept underlying all of biology and is supported by multiple forms of scientific evidence.

Kendall's proposal asked for the following revision (with changes shown here in italics):

> Evolution is [a] fundamental concept underlying all of biology and is supported by multiple forms of scientific evidence *and teachers should be permitted to engage students in a critical analysis of that evidence.*[49]

Kendall felt that adding "the scientific theory of" to the standards (Option B) was redundant and didn't change anything. She assured the board that she was not asking for creationism or intelligent design to be taught; it was fine to teach evolution and include it on tests. Instead, it was important to "emphasize the importance of critical thinking in the classroom." Other states—New Mexico, Pennsylvania, Minnesota, and South Carolina—had added the same phrase on academic freedom to their standards, she erroneously claimed. Changing that one sentence was the only thing she asked the board members to do.[50]

Rich Akin, chief executive officer of Physicians and Surgeons for Scientific Integrity, said his Clearwater-based organization was secular. Members of his group believed that current science had "disproved Darwinian macro-evolution." Akin felt that the science standards' treatment of evolution was dogmatic and based on a "consensus of science" that posed a barrier to scientific progress. The current definition of "evolution" in the standards was "frankly embarrassing to the state of Florida and does a disservice to the students," he said.[51]

The anti-evolution speakers repeatedly parried the pro-evolution speakers' claim that religion was a motivating factor, arguing that the issue was critical thinking, not religion. Kemple stated: "I'd like to be clear for our coalition of over twenty Florida organizations [that] this is not about religion; this is about academic integrity."[52] Later, Stemberger pointed out that nearly every pro-evolution speaker had brought faith into the debate, but not a single proponent of the Academic Freedom Proposal had mentioned any aspect of religion as motivation.

Throughout nearly all of the anti-evolutionists' speeches, the spotlight was on academic freedom. Patricia Weeks, chairwoman of the Baker County School Board, said that the science standards required critical thinking in all of the subjects except for evolution. "We are telling students that you are not allowed to use those skills to investigate or question. Instead you are to put those skills aside on this one issue of Darwinian belief," she said. Kemple and Stemberger said that issues like this one were prompting parents to pull their children out of public schools. "All across America there is already an exodus taking place from public schools," Kemple said. "Parents, tired of seeing their children indoctrinated with this type of dogmatic propaganda, are homeschooling them or enrolling them in charter and private schools."[53]

Other speakers on the academic freedom side were Lori Muller, a mother and certified teacher; Frank Cutting, who had submitted a "Minority Report" about evolution to the standards committee; David

Brackin, a certified Florida teacher for twenty-three years; Clifton Savoy, who had a Ph.D. in microbiology; and Robin Brown, a retired Polk County middle school science teacher.

However, many of their speakers recited talking points that creationists had used over the past few decades: that there are gaps in the fossil record, that some scientists have been fired for not believing Darwin, and that evolution is a house of cards on the verge of collapse. Stemberger said that scientists have never found an example of one living thing turning into another through the process of evolution. His example was: "Yet we look at the fossil record and we find rats and bats, but no transitional forms of 'rat-bats.'"[54]

World-Class Standards

After a short break, Mary Jane Tappen, executive director of the Office of Mathematics and Science, gave a presentation about the standards writing process. She was happy that there had been so much public interest in the standards, especially when compared to interest in other subjects' standards. As an example, she said that more than four hundred people had reviewed a standard about the structure of atoms. "I kidded the other day that we were thankful to get twelve people to look at our calculus standards," she said.[55]

She kept the focus of her presentation on the overall document and how it was a significant improvement over the 1996 version. In the previous version, eighth grade teachers had to cover one hundred science concepts in the school year. "If you approve these new benchmarks, they will get to teach—versus cover—forty benchmarks, which will allow them more time to allow students to experience science, ask about science and experiment with science versus just hear about science," she said.[56]

But then Tappen introduced Option B. Martinez seized that opportunity to go on the offensive. He grilled Commissioner Smith, rather than Tappen, about the timing and reason for its creation. When Martinez asked why Option B was conceived in the first place, Smith had trouble answering the question directly. Instead, he explained that a National Academy of Sciences publication that had been provided to the department of education identified evolution as both a fact and a theory. Additionally, some standards, such as plate tectonics, were referred to as theories, and some, such as evolution, were not. However, he never addressed Martinez's specific question.[57]

Martinez made it clear that he knew exactly why the changes had been made: "to placate those people that have concerns about the evolution standards." Smith responded that that wasn't the reason, but he clearly fumbled in his explanation, saying only that his communications with members of the committees had indicated that the phrase "theory of evolution" would be appropriate. Unsatisfied with Smith's responses, Martinez turned to Tappen and asked if the writers and framers of the original draft had been asked about Option B. He was told that an e-mail had been sent out to the group on Friday afternoon. Thirty-eight of them responded, with twenty-nine opposing Option B, two grudgingly accepting Option B if it was the only way to get the standards approved, and seven saying they were fine with Option B.[58]

Next, Martinez asked if Option B had been vetted by the National Academy of Sciences or the American Association for the Advancement of Science in the same way the original draft had. The answer was no. "Has any accepted mainstream science organization reviewed the Option B standards?" he asked. The answer was again no. "Then why are we even considering them, commissioner?" he asked.[59] The Option A standards had gone through a year-long creation process and been guided and reviewed by experts in order to produce "world-class standards," Martinez said. He went on to point out that in just a few days and with no expert input of the magnitude Option A went through; Option B was now before the board.

Martinez had managed to make this first attempt to throw out Option B, but then board member Raulerson stepped in to echo Smith's observation that the word "theory" was used in conjunction with some subjects in Option A but not in other places. She took that to mean that labeling a concept as a theory—she used the example of "cell theory" as it's written in Option A—"encourages more study." Adding the word "theory" to "evolution" would thus put it in the same category of being open to questioning. "At one point in time every educated person in the world believed it was flat," Raulerson said. "And what we need is more ideas out there that can be questioned and tested and tested. Not that they're wrong or that they're right, but that they're open for exploration."[60]

At this point, Callaway derailed the developing debate, pointing out that the board hadn't made a motion to approve the standards. So, the presentation that Martinez had interrupted was allowed to continue. After Tappen finished, Smith gave a short speech about the hard work everyone put into the standards and the huge improvement the proposed standards were over the old version. He recommended to the board that they should approve Option B because it would "help to clarify for our classroom teachers on how to address these concepts and bring better consistency to the document."[61]

Next, Fair clarified to the board members the choices before them and then waited for someone to make a motion to approve one. But Martinez first wanted to discuss the validity of Option B. Florida has strict laws that require open access to nearly all aspects of state government, he said. Citizens have rights to most government documents and access to government meetings. With this in mind, Martinez questioned whether the rushed nature of Option B violated any of these laws.

Callaway countered with the suggestion that first a motion needed to be made and seconded to approve Option A. Then, during discussion, any board member could propose to amend that document with Option B, or "choose to amend it in any other way with other additions to it. This may not be the only addition that's recommended in connection with the standards today." Shanahan pointed out that amendments such as the one Callaway was talking about come up all the time at board meetings without violating Florida law. She also said that she had brought up with department of education staff the issue of using the word "theory" with "evolution" back in December. Even if there was no Option B—which she had not known about until reading it in the newspaper—she had already planned to ask for the word "theory" to be added at the current meeting.

The board's legal counsel then gave the opinion that so far everything was perfectly legal and no aspect of public records law had been violated.[62]

"We Count This as a Victory"

A motion to approve Option B was eventually made and seconded. Discussion was officially launched, and Martinez once again took the lead. The efforts of experts over the course of a year had produced the world-class standards of Option A. Both Option B and the Academic Freedom Proposal were merely efforts to "water down" the standards "for reasons having nothing to do with science per se," he said. This was met with loud calls of "No!" from the audience. Undeterred, Martinez went on. "No matter how much the current strategy may have evolved over the last twenty years, the DNA is the same with its common ances- tor: creationism."[63] He pointed out that the critical thinking that other board members were calling for was already required in the "Nature of Science" section, so there was no need to single out evolution for special treatment.

Finally, Callaway interrupted. She asserted that despite her strong religious identity, her stance had nothing to do with religion, but was based on her extensive research. She lamented that the way evolution was presented in the standards made it too dogmatic, denying students their right to explore the issue for themselves as she had done in prepa- ration for the day's meeting. Her position was that Option B didn't ad- dress her concerns but that the Academic Freedom Proposal given to the board that morning was a simple and perfect solution. Thousands of people don't accept the theory of evolution, and kids needed to be made aware of that. We shouldn't try to hide the controversy that is out there, she said.[64]

As other board members then stated their opinions, the debate fi- nally took shape. Shanahan, Raulerson, and Taylor either ignored Cal- laway's academic freedom tactic or brushed it aside as not needed. They favored Option B. Desai didn't like Option B but seemed receptive to the Academic Freedom Proposal. Fair was the only person to completely stay out of the debate.

Although Callaway's push for the Academic Freedom Proposal never gained traction, the debate featured her and Martinez coming to ver- bal blows toward the end. Martinez insisted that Option B's whole in- tent was to single out and dilute evolution. "My question is, teaching the scientific theory of evolution as opposed to what other theory?" he asked. "The religious theory of evolution?" It was clear to him from all the correspondence he had received from people opposed to evolution that their motivations were religious in nature. One e-mail he quoted said, "I'm sorry to hear that your ancestors were monkeys. Mine were created in the image of God." Martinez explained: "We know what people are concerned about no matter how it is cloaked."[65]

"I take issue with the fact that you say you know where that's all coming from," Callaway responded. "I would like to see it separated away from the creation/intelligent design. There is not a person that I know of or that I heard of . . . who's advocating those at all." Martinez wouldn't be swayed, though, pressing the question of what alternative theory was out there. He compared it to teaching that the Earth revolves around the sun, but for the sake of critical thinking not ruling out the possibility that the sun goes around the Earth. Callaway countered by saying that kids needed to explore the issue because there are such great differences of opinion about evolution in the world. "If they come up with another theory, so be it," she said.[66]

She then tried to show that Martinez's insistence that there were no other theories was dogmatic and against critical thinking. He replied by saying that evolution is a fact and a fundamental concept in biology, which prompted groans from the audience. But Callaway said that there was still a debate on the subject and that students needed to know that. "Respectfully, Donna, it is not a point of debate or controversy in the mainstream scientific community," Martinez said, getting in the final jab as his supporters erupted with loud applause, drowning out whatever Callaway tried to say in response.[67] Fair then stepped in to scold the audience for its outburst.

While Martinez and Callaway cooled off, Raulerson restated some of her previous comments about the use of the word "theory" throughout the standards. Fair then asked if anyone had new information to discuss, and when there was no response, he finally called for a vote on Option B.[68] Taylor, Shanahan, and Raulerson voted yes. Ironically, Martinez voted no along with Callaway and Desai. Fair broke the tie with a yes vote. Florida now had a new set of science standards. A break was quickly called for, and the reporters lunged forward to grab interviews.

Martinez and Desai had voted no as a protest against Option B. They both believed that the original version, written and vetted by experts, was better, and that Option B watered down the standards for no valid scientific or educational reason. Callaway, in contrast, had voted no because she believed that the standards didn't go far enough in allowing academic freedom to question evolution in the classroom.[69]

Florida Citizens for Science's reaction was celebratory, even with "scientific theory" squeezed in at the last minute. "It is perfectly fine with us," said Jonathan Smith. "We count this as a victory for all the children for the next 10 or 12 years who go through our education system."[70]

The vote had been watched closely by stakeholders outside of Florida, too. "We see [the decision] as a victory of high-quality science education," Gerry Wheeler, executive director of the National Science Teachers Association, told *U.S. News and World Report.* "My only concern is that 'theory' might be used as a wedge to start teaching controversy where there is no controversy."[71]

Taking a much harsher stance was Paul Gross, who had reviewed state science standards for the Thomas B. Fordham Institute. He didn't like the compromise, since it gave in to people who believed that "some other way of knowing" is just as valid as science. "This may well be good politics," he said, "but it is not serious; and, so far as excellent science education is concerned, it is dishonest."[72]

Stemberger said that the newly approved standards failed on the academic freedom front. "This is a sad day in Florida, when a handful of religious Darwinists can hijack the curriculum framing process and push their ideological agenda at the expense of the education [of] our children," he said. The Discovery Institute's Casey Luskin agreed. "Those board members were tricked into a false compromise," he wrote. "Inserting the word 'scientific theory' before the word 'evolution' is a meaningless and impotent change that will do absolutely nothing to actually inform students about the scientific problems with evolution."[73]

Stemberger, Kendall, and Kemple all announced that their next stop would be the state legislature. Kemple told the *Tampa Tribune* that "about a half-dozen lawmakers" were interested in incorporating the Academic Freedom Proposal's changes into the standards. "I don't want to give up the fight yet," Kendall told the *Florida Baptist Witness.* The

Witness confirmed that House Speaker Marco Rubio felt sure that there were enough votes in his chamber to pass a bill addressing academic freedom. Rubio said he was concerned that what children learned at home could be "mocked and derided and undone" in the public schools. "And for me, personally, I don't want a school system that teaches kids that what they're learning at home is wrong," he said.[74]

In the same issue, the *Witness* published a column by Callaway. She referred to the evolution debate as a battle and said that Florida's students were the losers. "They lost some privileges and they lost some rights," she said. She recalled that she had not known that evolution was even in the standards until she read about it in a newspaper. This motivated her to request "an immediate copy of the new standards with every mention of evolution tagged. The notebook arrived and the bright yellow tags were many." All of the evolution references were "setting off alarms" for Callaway. "I contacted the executive editor of this publication because I felt that this standard needed to get out to Christian parents, children and churches throughout the state," she said.[75]

Callaway went on to praise those who opposed the teaching of evolution as being calm and focused on students' rights, which "confused their opponents who expected a religious battle." She said, "I left the SBOE meeting emotionally drained but reaffirmed by the love for children and the respect for others that I saw in those who hold beliefs with which I can identify."[76]

10

"Who Gets to Decide What Is Science?"

Politics is "not for pansies," Ronda Storms said when she was a Hillsborough County commissioner in 2004. She had no problem speaking her mind when it came to her strong Christian conservative convictions. The rural citizens she represented appreciated how she always put them first and controlled government spending through fights against tax and fee hikes. Those on the opposite side of the fence felt she was needlessly combative and disrespectful and that she talked down to those who didn't support her.[1]

Storms was a former high school English teacher who then quit to obtain a law degree. After two years in a legal practice she jumped into local politics. During her tenure as commissioner she became famous for battling adult entertainment businesses and banning Hillsborough County from officially recognizing gay pride events. Then in 2006 she won a seat in the state Senate where it didn't take her long to stand out. She had announced her candidacy on a Christian talk-radio station where she unabashedly discussed how her religion guided her political career.[2]

Her words were converted into action when she filed a bill in 2008 to tax adult businesses—strip clubs and escort services—with the money going to low-income nursing home residents. Another 2008 bill of hers

Sen. Ronda Storms (R-Valrico) filed in 2008 the "Academic Freedom Act," which allows public school teachers to objectively present in their classrooms scientific information relevant to the full range of scientific views regarding biological or chemical evolution without fear of reprisal. (Courtesy Florida Senate, official photo.)

would have created specialty vehicle license plates featuring a cross and the words "I believe" with the proceeds going to a sectarian private school.[3] Both bills grabbed the public's attention, but neither passed.

However, the bill that really propelled Storms into the media spotlight that year was the "Academic Freedom Act." A mere ten days after the Florida Board of Education approved new state science standards that included the teaching of evolution, Storms filed the bill that antievolution activists had hinted at throughout the previous few months. The stated purpose of the bill was to protect teachers and students: "The Legislature finds that in many instances educators have experienced or feared discipline, discrimination, or other adverse consequences as a result of presenting the full range of scientific views regarding chemical and biological evolution."

The meat of Senate Bill 2692 was contained in two paragraphs:

A public school teacher in the state's K–12 school system may not be disciplined, denied tenure, terminated, or otherwise discriminated against for objectively presenting scientific information relevant to the full range of scientific views regarding biological or chemical evolution in connection with teaching any prescribed curriculum regarding chemical or biological origins.

Public school students in the state's K–12 school system may be evaluated based upon their understanding of course materials, but may not be penalized in any way because he or she subscribes to a particular position or view regarding biological or chemical evolution.[4]

The issue of religion had persistently clouded the recent battle over the teaching of evolution, so the bill also included this paragraph:

This section shall not be construed to promote any religious doctrine, promote discrimination for or against a particular set of religious beliefs, or promote discrimination for or against religion or nonreligion.[5]

This bill was not a Storms original, though. It was a modified and localized version of one promoted by the Discovery Institute. This fact would cause problems for Storms later when other senators grilled her for information on what was meant by "full range of scientific views." John Stemberger and Terry Kemple, two conservative activists who were at the forefront of opposition to evolution in the state science standards, were credited with helping Storms draft and file the bill.[6]

Four days after Storms introduced her bill, Rep. D. Alan Hays filed the same bill in the other legislative chamber. The semi-retired dentist

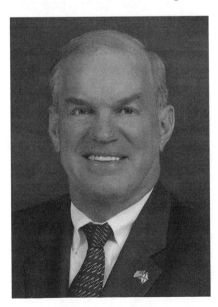

Rep. (and later senator) D. Alan Hays (R-Umatilla) filed a companion bill in the House to the 2008 "Academic Freedom Act" filed in the Senate, but he significantly altered his bill later in that legislative session. (Courtesy Florida Senate, official photo.)

had been a reliable conservative voice in the House since his election in 2004. The Christian Coalition of Florida gave him their Faith and Family Award every year of his service. During his first election campaign, one of his biggest issues was education, including returning some religious values to the schools. "The idea of separation of church and state today is an aberration of what our forefathers intended," he said.[7]

In conjunction with the bills' filing, Kemple issued a press release with his strongest anti-evolution language yet. Evolution is a "sacred cow" that all of academia must march in "lock step" with or face career-debilitating persecution, he said. He claimed that any teacher who wanted to challenge the concept of evolution "did so at the risk of his or her employment."[8] This became the main theme in arguments supporting the bills. Teachers merely wanted protection so they could freely investigate the strengths and weaknesses of evolution with their students. Otherwise, teachers felt they would be ostracized by their science departments or even face more outright reprisals, according to Storms, Kemple, and Stemberger.

But just like the bills' language, the overall theme came from out-of-state influences. The Discovery Institute quickly took exception to Florida news media's reports that the bills would allow alternative theories, such as intelligent design, to be taught. The Discovery Institute's director of communications, Robert Crowther, said that teachers definitely should teach evolution, "But if a teacher also presents some of the scientific evidence that challenges the theory, they should not be reprimanded."[9]

"Intellectual Dishonesty"

The *St. Petersburg Times* asked Kemple and Hays for examples of teacher persecution in Florida, and they came up with two. Veteran teachers David Brackin and Robin Brown were among the ten anti-evolution speakers who had addressed the board of education in February. Brackin, an Orange County public middle school science teacher, told the *Florida Baptist Witness*: "As a young teacher many years ago I was brought into the office and told not to teach religion. I wasn't teaching religion. I showed some weaknesses in evolution." Brackin left that school, and he revealed to the *St. Petersburg Times* that he currently worked for a principal who "believes in academic freedom."[10]

Although there was scant evidence of a persecution problem, Hays felt the protection was needed. It didn't matter "if it's four teachers in the whole state who were harassed. . . . We want the teachers to be able to teach the full scope (of evolution) without fear." Some legislators, though, saw the bill as nothing more than a rerun of the 1925 Scopes "Monkey Trial." "I never thought I'd be in the Florida Senate in the 21st century, still having the same debate about evolution," said Senate Minority Leader Steve Geller. Rep. Joe Pickens, chairman of the House Schools and Learning Council, didn't feel that it was lawmakers' place to jump into the debate. "My expectation is that there isn't a great deal of appetite to go in and undo what the state Board of Education did in their purview, under their authority," he said.[11]

The bill's supporters, however, kept their focus on having the freedom to expose the flaws in evolution. "You and I both know there are holes in Darwin's theory," Hays told a reporter. "No one yet has found a half-animal of this or a half-insect of that. And they certainly haven't found any half ape and half man." To prove his point that evolution was being presented dogmatically, Hays invited his fellow House members to an "exclusive pre-screening" of the documentary *Expelled: No Intelligence Allowed*. The movie starred Ben Stein, best known for his role in the movie *Ferris Bueller's Day Off* and his work as a former presidential speechwriter and television game-show host. Hays's invitation described the movie: "*Expelled: No Intelligence Allowed*, follows Ben Stein on his journey around the globe where he discovers that scientists, educators and philosophers are being persecuted because they dare to go against the theory of evolution."[12]

The March 12 screening was strictly for lawmakers, their spouses, and legislative aides. Everyone else, including the media, was banned unless they signed waivers promising not to reveal anything about the film. "It's kind of an irony: The public is expelled from a movie called *Expelled*," said House Democratic leader Dan Gelber. Because Florida has strict open-meeting laws that are triggered anytime two or more elected officials meet to discuss government business, Gelber questioned the screening's legality. The House general counsel assured Gelber that as long as attendees didn't talk about the issue with each other the event would pass muster. Larry Spalding at the Florida ACLU was skeptical. "I guess if they do what they say they're going to do, it's not an open meeting," he said. "If they do what everyone thinks they're going to do, then they're violating open meeting laws."[13]

The day of the screening, supporters of the bills held a press conference in the state capitol to promote the film and drive home to reporters the dire need for the legislation. Stemberger kicked the event off with an impassioned assault on his opponents. He recounted the many causes he had worked for over the years and expressed his amazement at the antagonism he now faced challenging evolution. "I've never ever in two decades of that social activism seen more hostility and more mischaracterization, more intellectual dishonesty and more just bad faith from our opponents than on this issue of challenging Darwinism," he said.[14]

Stemberger was joined by many of the same people who helped oppose approval of the science standards: Robin Brown, Rich Akin from Physicians and Surgeons for Scientific Integrity, and retired engineer Fred Cutting. They all repeated the point that teachers need protection from "discrimination, hostility and intolerance" from "the religious, hardcore Darwinists."[15] Several times the speakers implored reporters to focus on the bill's text and stop bringing the perceived motives and personal beliefs of the bill's supporters into their stories.

The stars of the press conference were Stein, Hays, and Casey Luskin from the Discovery Institute. Luskin claimed that opponents were promoting censorship and "fear mongering." Teachers in other states have lost their jobs because they dared to challenge evolution, he said. Stein echoed the other speakers' assertion that there was no intention to insert intelligent design into the curriculum. The issue was freedom of speech, and Stein compared the plight of teachers who want to question evolution to the civil rights movement. "The Darwinists, and the neo-Darwinists in particular, have become more Darwinist than Darwin in the sense that they will simply not permit any questioning of Darwin, whereas Darwin encouraged—applauded—questioning of his theories and beliefs," Stein said.[16]

After thirty minutes of speeches, the floor was opened to questions. Reporters demanded to know what qualified as the "scientific information" the bill called for and who would make that determination. Stein suggested that teachers could make that decision themselves, but no one else offered a clear answer. Luskin was caught in an apparently uncomfortable tug of war. His employer, the Discovery Institute, was the most prominent promoter of intelligent design in the country, but the bill's supporters were trying to distance themselves from the concept. Under pressure from reporters, Luskin admitted that intelligent design might qualify as scientific information. "In my personal opinion, I think it does," Luskin said. "But the intent of this bill is not to settle that question."[17] He went on to assure reporters that the Discovery Institute wasn't looking to have intelligent design "mandated" in any schools.

The Florida ACLU's response to the press conference seized on Luskin's admission. Executive Director Howard Simon told the *Miami Herald* that the bill was just part of an overall effort to get religion into science classes. "The strategy is this: Let's call intelligent design scientific information, and let's make sure that teachers can teach that scientific information," Simon said. In an ACLU press release, Simon had harsh words for lawmakers: "Allowing schools to masquerade Intelligent Design as science would be a blunder and an embarrassment for the Florida Legislature."[18]

Reporters were on hand the night of the screening even though they were barred from entering the theater. Despite all the initial hoopla, it was just an uneventful night at the movies. About one hundred people attended, but only a handful of them were lawmakers.[19]

"The Pot Is Still Boiling"

Intelligent design haunted Hays and Storms from this point onward. Storms, who wasn't associated with the press conference or movie, was quoted on March 11 in the *St. Petersburg Times* education blog as saying: "Under this bill, if you have a teacher who is pro-evolution and every student is intelligent design . . . that teacher is safe to teach that as a theory." But six days later she told St. Petersburg TV station WTSP that her bill "does not promote creative/creation science and it also does not promote intelligent design."[20]

Despite the constant assurances from Hays and Storms that these bills were just about basic academic freedom, they took quite a bit of heat in the media. *South Florida Sun-Sentinel* columnist Michael Mayo felt certain that the proposed legislation was "a thinly veiled attempt" to get religion into public schools. "The bill is written with so much mumbo jumbo and wiggle room, you wonder what the true motives are," he wrote. An editor at the *Tallahassee Democrat* predicted chaos in the classroom: "If the full Legislature goes along with this galling intervention into daily classroom instruction, teachers will be able to teach whatever they want regardless of state standards, and teenagers will be endlessly debating the specifics of things they cannot possibly understand because they've never been taught solid, well-considered and agreed-upon instructional materials."[21]

As the newly introduced bills buzzed with building controversy, the obvious question on everyone's minds was if the bills would gain any traction in the legislature. The chairman of the Senate's Education Pre-K–12 Committee, Don Gaetz, indicated that the door was open. "I had hoped the Board of Education would resolve the scientific standards, but obviously the pot is still boiling," said the former school superintendent. In both chambers, Republicans, most of whom favored the bills, far outnumbered Democrats, most of whom opposed the bills. If the bills gained momentum, there was little that Democrats could do. Despite this challenge, the minority party still jabbed when they could. "The State Board of Education has addressed it and that's the end of the debate," Gelber said. "This is a solution in search of a problem."[22]

Upset that their hard work was yet again under attack, thirty-seven of the science standards' writers and framers issued a press release on March 17. Evolution is "the only explanation of the development and diversity of life" that is based on strong scientific evidence, they pointed out, calling the bills "subterfuge for injecting religious beliefs held by some into the science classroom." Florida Citizens for Science's Jonathan Smith called the bills "blatantly religious" and pointed out that they addressed only evolution. "Why not academic freedom to teach the stork theories in sex education?" he asked.[23]

Once a bill is filed in the legislature and officially introduced via a "first reading" it is then typically sent to appropriate committees for review. The handful of lawmakers assigned to those committees can debate, make changes to the bill, hear from the public, and finally kill the bill or approve the bill for consideration by the full House or Senate. It can be a tough gauntlet to pass through, and many bills are stalled or discarded in this process. Storms's and Hays's bills were assigned to the Education Pre-K–12 and Judiciary Committees in the Senate and the Schools and Learning Council in the House, respectively.

"The Full Range of Scientific Views"

Before a committee considers a bill, staff personnel conduct research on the subject on behalf of the legislators. The resulting analysis documents are then distributed to committee members before the meetings. The reports prepared for the three committees raised a number of interesting points. All three reports emphasized that the new science standards already had a "Nature of Science" section that was intended to "encourage teachers and students to discuss the full range of scientific evidence related to all science, including evolution." They also noted that the department of education never had a case of a teacher or student claiming that he or she had been "discriminated against based on their science teaching or science course work."[24]

Before the analysis was conducted, Storms tweaked her bill, changing its title from "Academic Freedom Act" to "Evolution Academic Freedom Act." She also added a paragraph: "As used in this section, the term 'scientific information' means germane current facts, data, and peer-reviewed research specific to the topic of chemical and biological evolution as prescribed in Florida's Science Standards." Instead of soothing any concerns about what could be used to challenge evolution in the classroom, the new paragraph raised the analysts' eyebrows. Who would determine if information presented by the teacher met that definition? The Senate education committee analyst wrote: "The administration and the teacher may have quite different views on the objectiveness of the information presented." Analysts were also wary about the rights provided to students by the bill. Would all students be tested and evaluated equally, or would those with alternate opinions about evolution have separate accommodations? "This ambiguity may create unanticipated problems with student evaluation and grading in science classes," wrote the Senate judiciary committee analyst.[25]

The House education council analysis was blunt in its assessment of the constitutionality of Hays's bill. The bill would protect teachers who wanted to present "the full range of scientific views" on evolution, the analyst wrote, but "teachers are not granted such a 'right' in any other subject area. This fact raises issues concerning the underlying intent of the bill." The analysis warned that teachers, school administrators, and district staff might all have differing opinions on what is scientific, leading to "frequent and challenging" clashes that "may increase litigation expenses for the school district."[26]

A "Hostile Environment"

Storms got her first chance to sell her bill to her fellow senators on March 26 at the Education Pre-K–12 Committee meeting. She told her peers that her bill did nothing more than allow teachers to tell students about the flaws in evolutionary theory. Religious belief had nothing to do with the bill, she said. According to Storms, some teachers were being discriminated against in Florida. They might not face severe punishments like dismissal, but extra bathroom and lunch room duties could make a teacher's job difficult. "There are myriad ways to punish teachers for stepping off the reservation," she said.[27]

Committee member Larcenia Bullard, a Democrat, was concerned about exposing kindergartners to evolution, saying "that may be brainwashing." She also related a bad experience from her college days. Her professor had warned students that unless they answered test questions on evolution according to how he taught the information, they would fail. Bullard said she defied him by answering the questions with Bible verses from Genesis. A few citizens in attendance then shared their views. Courtenay Strickland, representing the ACLU of Florida, warned the senators that a "massive" lawsuit was waiting if the bill passed and a teacher used it as cover to teach religion. "You cannot simply call a religious belief scientific information and thus open the door to teaching it in our scientific classrooms," she said. David Brackin, the teacher who had said he faced persecution for questioning evolution in the classroom, was at the meeting but declined to speak before the senators. However, he told a reporter, "I don't want to say evolution is fact. I don't want to say intelligent design is a fact. I want my kids to question. I want to be able to do that without being dragged into the office."[28]

Only five of the seven committee members were on hand to vote, but it's unlikely that the absentees would have changed the outcome. The final vote was 4–1, allowing the bill to clear its first hurdle with ease. Sen. Ted Deutch had cast the lone dissenting vote. "What we heard today was the suggestion that there are people of faith who have some objections and they're not permitted to raise them," he said.[29]

There was a lingering doubt, voiced by Deutch and the bills' analysis documents, that there were actual victims in Florida who would benefit from this legislation. Brackin, the veteran teacher who claimed to have faced discrimination early in his career, worked on remedying that perception. He issued an alert through Christian Educators Association International and posted the call to action on an Internet message board at injesus.com. Because of the approval of the new science standards, many teachers were now forced to "teach the lie of evolution as though it is a proven fact," he wrote. The alert asked for any public school teachers who worked in a school with a "hostile environment" toward those who disagree with Darwinism to contact Brackin. Florida Family Action put out their own alert encouraging supporters to inundate lawmakers with phone calls, e-mails, faxes, and letters. "The bill will help to defuse the public controversy caused by the one-sided, pro-Darwinian views that are inherent in the new science standards," the alert said.[30]

The bill's next stop in the Senate was the judiciary committee, on April 8. Storms introduced her bill to the committee members, saying it would protect teachers who present information "critical of Darwinism or evolution." This protection was needed because the new science standards presented evolution "in a dogmatic way and assumes its validity without critical thought or examination."[31] She assured the senators that the bill made the standards stronger and didn't mandate the introduction of creationism or intelligent design.

Deutch, who had cast the only no vote in the bill's other Senate committee meeting, was also on this committee. He peppered Storms with questions, specifically attacking her claim that many teachers felt "muzzled" or "intimidated." The back-and-forth grew testy as Deutch stated that there was no evidence of unfair treatment. Storms countered that Deutch was well aware that six people had been prepared to offer their testimony at a previous committee meeting but couldn't because of time limits. However, Deutch still wasn't convinced that "this whole parade of horribles has actually played out across the state."[32]

Next, Deutch and Sen. Jeremy Ring asked whether similar bills had been considered in other states or on the federal level. Storms quipped that she had limited her analysis to Florida law. But a committee staff attorney outlined a few examples for the senators. Annoyed at the line of questioning, Sen. Don Gaetz supported Storms by pointing out that the staff's bill analysis stated that creationism and intelligent design weren't mentioned in the bill. Several times Storms had to fend off the grilling over creationism and intelligent design. "It does not say it. It does not imply it. And furthermore it expressly declines to promote religious doctrine," she said at one point.[33]

During a short public-comment period, Florida Family Action Network's vice president, Nathan Dunn, addressed Deutch's doubts about oppressed teachers. He read from letters sent to his organization from teachers who supported Storms's bill. One teacher wrote: "I have been criticized for offering scientific evidence that goes against evolution. There's always a dark cloud looming in the back of my mind even though the evidence I provide is purely scientific. I shouldn't have to teach under these restraints." Another teacher wrote: "My school is a majorly hostile environment for criticism of evolution and I can't wait to transfer. I greatly fear reprisal from administration as I have already been completely shunned and ostracized by my science department."[34]

Becky Steele of the ACLU was direct and scathing as she asked, "Who gets to decide what is science?" The definition of science in the bill was very broad and could be an invitation to teachers to discuss intelligent design in class because they think it's science, Steele said. She accused Storms of being evasive by answering questions "very carefully and very narrowly" to show that the teaching of intelligent design wasn't mandated. "But she did not go so far as to say that this bill would forbid a teacher from presenting intelligent design teaching in the science schools."[35]

Steele was followed by Mary Bahr from Florida Citizens for Science, a veteran science teacher who had served on the science standards writing committee. She pointed out that the standards already had a clear "Nature of Science" section, but that clarity would suffer if Storms's bill passed. "So my question as a classroom teacher is what does this bill, if it's passed, direct me to do that I would not already do under the new standards?"[36] This went unanswered.

Just before the committee's vote, the senators' debate concluded just as tensely as it had started. Sen. Geller said that he could probably find a scientist willing to testify that the world is flat, but that wouldn't make it so. He was confident that Storms's bill was about religion in schools. Sen. Dan Webster countered that the bill just allows teachers to point out flaws, which every theory has. He didn't want teachers giving "canned speech" lessons.[37] The bill passed 7–3. The full Senate could now debate and vote on it if the Senate leadership decided to make room in the tight schedule.

A Different Persuasion

Meanwhile, in the House, Hays's bill made its only committee stop on April 11, where it underwent a radical change. The Schools and Learning Council spent about an hour on debate and citizens' comments, but they didn't discuss the seven-paragraph bill originally filed. Instead, Hays introduced a "strike-all" amendment that erased his original language and replaced it. Now, instead of creating a new statute, Hays's bill merely amended an existing statute titled "Required Instruction" that listed several specific subjects that school districts must ensure are covered, such as "flag education," "women's contributions to the United States," and "the history and content of the Declaration of Independence." Hays's bill would add another item to the list: "a thorough presentation and critical analysis of the scientific theory of evolution."[38] Hays explained that there might have been constitutional issues with the first version that the simpler revision avoided. "The amendment was offered because we felt that the original language had a greater risk of opening the door for religious teaching," he said.[39]

In one lawmaker's opinion, the new bill still posed a legal hazard. Democrat Martin Kiar proposed an amendment to the bill that would delete the words "critical analysis," because they could be interpreted to include creationism, which the Supreme Court had ruled unconstitutional. Other committee members disagreed and accused Kiar of trying to stifle critical thinking in classrooms. Undeterred, Kiar gave a melodramatic speech claiming that a vote against his amendment was a vote against the Constitution and the Supreme Court. The other representatives gave him a good-natured ribbing about his passion, but his amendment failed.[40]

Kiar then asked Hays if his bill would mandate or at least allow the teaching of creationism. Again and again Hays denied it. "You're doing something so many people so frequently do," Hays said. "They instantly jump and raise a big red flag and say, 'Holy cow, you're teaching religion.' And nothing could be further from the truth." Kiar wouldn't give up, though.

"Do you think that if this case is challenged in court that it could survive a constitutional challenge over the establishment clause of the United States Constitution?" Kiar asked.

"Sure," Hays answered.

"Why?"

"'Cause my counsel tells me it will. You just happen to be of a different persuasion."

"Who is your counsel?"

"That's for me to know," Hays said. He, Kiar, and the other lawmakers laughed at the answer, which ended Kiar's questioning.

Next, Rep. Shelley Vana, another Democrat, said she didn't understand why the bill singled out evolution for special treatment. Why not cell theory or other science concepts? Hays said that he heard from teachers who were afraid to offer a critical analysis of evolution, but he wasn't aware of similar concerns in other subjects. Vana said that she had taught evolution as a high school science teacher for several years, and she was concerned about the bill because as a teacher she didn't know what she would have to say to her students that would qualify as the mandated "critical analysis." Hays responded: "I would suggest that you just keep on teaching the way you have been teaching and you will probably be just fine."[41]

Hays then tried to elaborate on the bill's purpose. "The whole idea is to make sure that the teachers feel comfortable and insulated, if you will, from criticism that because you're not teaching the theory of evolution as THE origin of the species, as the basis for life, et cetera." The sentence was left hanging for a moment as Hays fumbled for the right words. He continued: "They need to be able to teach their students to evaluate the theory in an analytical way without fear of retribution."[42]

Next, seven people took advantage of the public-comment portion of the meeting. Among them was activist mom Kim Kendall, who related how she was helping one of her children with homework about the solar system when the child told her that Pluto was no longer a planet. She used the story as an example of science sometimes getting something wrong and teachers needing to bring that new information into the classroom instead of waiting for the next standards or curriculum update years later. Evidence countering evolution, such as the lack of transitional fossils, needs to be presented to students, she said. Although Kendall didn't have official authority in state government, Vana asked her if the bill would override the board of education's approval of the science standards. Kendall said that it actually matched what the board had intended. She claimed that the Academic Freedom Proposal that she had presented to the board had not been enacted only because of "parliamentary procedure issues."[43]

Other citizens then addressed the committee. A couple of men practically preached to the committee about how taking God out of schools had caused the nation to fall. A few other speakers had served on the science standards committees and explained that the bill was completely unnecessary. Strickland, from the Florida ACLU, was the final speaker. As the daughter of a Baptist minister and high school science teacher, she said, "these are not abstract questions." The bill came from the Discovery Institute, and its purpose was to wedge creationism into the schools, she said. "It opens the door to the teaching of religious beliefs as science in the classroom. Calling something science doesn't necessarily make it science. And unfortunately, under this particular legislation a teacher could choose to present something that was even in contradiction to the standards or the curriculum itself and argue that that was in fact critical analysis."[44]

Before the committee voted, Hays made a closing statement: "This bill will allow teachers to teach their students how to think rather than what to think."[45] His bill passed on a strictly party-line vote of 7–4. Just like the Senate bill, the House bill was now cleared for consideration by the full chamber.

"An Abuse of Position"

As the bills moved through the legislature, Florida Citizens for Science called for a press conference and roundtable discussion in Tallahassee on April 14. Nine prominent science and education organizations sponsored the event, including the National Association of Biology Teachers and the National Science Teachers Association. Speakers included Dr. Harold Kroto, 1996 Nobel Prize winner in chemistry; Vic Walczak, ACLU of Pennsylvania's legal director, who played a prominent role in the federal court case challenging the teaching of intelligent design in Dover public schools; and Dr. Ann Lunsden, a Florida State University biology professor and past president of the National Association of Biology Teachers. A press release announcing the event set the tone for the discussion: "This bill takes control of the curriculum away from education professionals. It is a red herring designed to suggest that discrimination exists where it does not, and to draw attention away from the subject matter of creationism. No law is required to authorize the teaching of scientific fact."[46]

Kroto took the lead in bashing the bills. "As far as I'm concerned, it's an abuse of position not to teach science correctly to children," he said. He explained that a few years prior Florida had spent millions of dollars to entice the Scripps Research Institute, a biomedical science organization, to establish a facility in the state. Based on comments made by colleagues in other countries, Kroto said, these bills were making Florida a laughingstock. "If I was a young scientist asked to come to work at Scripps in Florida, I'd think very carefully whether I'd want my children to go to school here," he said.[47]

Contrasting what was said at the roundtable meeting, Storms was just as avid in stating her case for the bills to the *Florida Baptist Witness*. She told the newspaper that ever since the Senate committees passed her bill she had been flooded with insulting e-mails. "By and large most of the people who have contacted me [against the bill] immediately launch into personal attacks, religious attacks—just being offensive about matters of religion—and then making personal attacks," she said.[48] But the opposition motivated her to keep going. These were exactly the types of attacks her bill was intended to address.

Each bill still had a few hurdles to overcome now that they were approved by all the assigned committees. First, there would be a "second reading" during which lawmakers in each chamber could debate and propose amendments prior to voting. Next, each bill would face a "third reading" where there would be more debate and then the final voting. Then, if both chambers approved their bills, any differences between the two would need to be ironed out in joint committees. All of this would have to be completed within the busy sixty-day session. Provided the bills could survive the entire process, they would become a single bill and would go to the governor.

"Asked and Answered"

Storms's bill was the first to face debate. During the second reading, on April 17, Storms dueled for about half an hour with Senators Deutch, Geller, and Nan Rich. First, Deutch proposed an amendment that was designed to determine how far Storms was willing to go to protect academic freedom. The proposed amendment said, in part:

> A public school teacher in the state's K–12 school system may not be disciplined, denied tenure, terminated, or otherwise discriminated against for objectively presenting scientific information relevant to the full range of scientific views regarding biological or chemical evolution and comprehensive sexual education that is age-appropriate and factual in connection with teaching any prescribed curriculum regarding chemical or biological evolution and any prescribed abstinence-only curriculum regarding human sexuality, respectively.[49]

Deutch explained that his amendment "expands the protection afforded to the teachers in our state" provided by Storms's bill and gives students freedom to ask questions that may help them understand how to prevent an unwanted pregnancy or prevent a disease. He said it "simply requires that those questions be answered, and that those students not be any more afraid of asking those questions than those students who may question evolution."[50]

"I'm concerned about prematurely deflowering kindergartners and first and second graders," Storms responded.[51] Although Deutch pointed out that his amendment specifically stated the information would have to be "age appropriate," the amendment failed on a voice vote.

With no other amendments to consider, debate turned to Storms's original bill. Geller pressed her for concrete examples of teachers who had faced discrimination. He referenced her bill's introduction, which stated that there were "many instances" of teachers facing "adverse consequences." Storms replied that many teachers are fearful and so feel they can't come forward. That's precisely what makes the bill so important. It gives these teachers the academic freedom they need, she said.[52]

Next, Rich took up the issue of teaching intelligent design. She asked if a science teacher would be disciplined for teaching it. Storms answered by reading excerpts from her bill about permitting the "full range of scientific views" and then said that the bill did not allow the teaching of religious doctrine. "It specifically says that. I don't know how else to say it," she said. Rich wasn't satisfied, though. She restated her question, pointedly asking that if a teacher taught intelligent design "then you are saying a teacher would not be allowed to do that." Storms turned to Senate president Ken Pruitt, who was presiding over the debate, and stated "asked and answered," which cut off Rich's questioning.[53]

Deutch then tried a different tactic. Storms had previously stated that she wanted to "hold the theory up so the theory can be questioned," and he wanted to know if that was true only in the case of evolution or if other theories—relativity, cell theory, atomic theory—should be held to the same standard. Storms didn't answer the question directly, but explained that students need to know and be tested on "what everybody commonly understands" but it was also acceptable to say "but there are people who believe different things."[54]

Sen. Larcenia Bullard posed the next question, which added an interesting twist to the debate. She wanted to know "how to prevent the abuses" of a child who comes from a Christian home where the family believes in creationism but then hears in the public schools that "there's another concept and that is Darwinism." She was worried that a student might become "so enthused by the teaching of Darwinism that this child now changes his or her total perspective and then it becomes another problem in our society with children who are trying to build families." How would children be protected from the abuses of a teacher who went too far? This time, Storms gave a direct answer. "Under today's law, the scenario that you described, without my bill, the scenario as you described can occur today."[55]

Toward the end of the debate, Geller picked up where Rich had been cut off during her questions about intelligent design. He told the Senate president that Rich's questions had not been answered to his satisfaction. Storms had previously fended off questions about intelligent design during debate by saying that teaching religion was not allowed under her bill, but that still left Geller wanting a definitive answer. "Can you teach intelligent design in the classroom under your bill?" he asked. Storms turned back to her bill and again read excerpts from it. Then she said, "The bottom line is if it is not scientifically based and if it is not scientifically relevant, the answer is no. If it is, the answer is yes."[56]

Geller accused Storms of dodging the question by giving three non-answers so far. "So, I'm going to ask again for the last time and if you don't answer this time I think we're all going to know what the answer is." It's a yes or no question, he insisted. "Under this bill, could you teach intelligent design in the classroom?" Once again looking down at her bill, Storms said that it provided a checklist that helps determine if information is scientifically based and relevant. "We don't know what science will do in the future," she said.[57]

Geller asked to follow up, and the Senate president cautioned him to not ask the same question again. "You haven't been willing to give us an answer on this so let me ask you your personal view," Geller said. The Senate president interrupted, telling Storms that she didn't have to answer the question if she didn't want to. Geller was then allowed to continue. "You're a teacher. You're the sponsor of this legislation so I think it is relevant to us what you think because you're the sponsor of the legislation. Do you believe that intelligent design meets the criteria in your bill?" Storms replied, "I absolutely believe that evolution should be taught in public schools. I also believe that we should teach the full range of critical analysis of evolution. That's what my personal belief is. I'm a teacher; however, I'm more inclined to teach Shakespeare since I'm an English educator and not a science educator. So, that's my answer."[58]

The Senate president asked if there were any more questions; since there were none, he moved on to the next bill. Geller later expressed his frustration to reporters, saying, "We could have stuck bamboo shoots under her fingernails and she wouldn't have answered."[59]

"Career Suicide"

A week later, Geller got one final opportunity to corner Storms. The third reading and the full Senate vote were held on April 23. However, because Hays's bill in the House had veered off in a different direction, Storms decided to do a bit of maneuvering, too. She introduced a strike-all amendment to mirror what Hays had done in the House. She told the senators that the single-sentence House version was very simple in that it merely provided for a thorough presentation and critical analysis of evolution. Storms had a problem, though. Because this switch wasn't proposed until the third reading, it would require two-thirds of the senators to vote in favor. The hostile reception the amendment stirred up indicated that that wasn't going to happen.

Geller was the first to question Storms about the amendment. Would the amendment mandate or permit the teaching of intelligent design? Following her earlier tactic, Storms read from the amendment, adding this time that critical analysis "would be the full range of debate on evolution." Geller refused to be put off. He asked again and requested that Storms give a yes or no answer, or at least give an answer that didn't entail reading from the bill. If Storms refused to answer directly, then they would all have to draw their own conclusions, he said. "It's my experience that people are going to draw their own conclusions whether or not I answer in the way they want me to answer," Storms replied. The reason why she kept referring back to the bill was that its "plain language" was the best response. "I know you want me to deviate from it, but I can't do it because it's not appropriate."[60]

Deutch questioned whether the amendment actually required all science teachers to give a thorough presentation and critical analysis when they taught evolution. Storms assured Deutch that he had read the amendment correctly and, as a matter of fact, that critical analysis should be the goal in any subject. Deutch seemed genuinely shocked and stated that it "makes a mockery" of the original bill. "This amendment doesn't protect teachers," he continued. "This amendment requires teachers to offer alternative theories to evolution. So, for those teachers who choose not to believe that the science class is the place to offer alternative theories of evolution, should they choose not to offer alternative theories, they can then face all of the punishments that the underlying bill was meant to prevent."[61]

A voice vote then convincingly killed the amendment.

Next, debate began on Storms's original bill. The bill's supporters advocated for free speech, such as when Gaetz said, "There is nothing wrong with inquiry. There is nothing wrong with debate. There is nothing wrong with discussion. And that's what this bill does and that's why it should be supported." Stephen Wise told his fellow senators that he had watched Stein's *Expelled* movie, which revealed how some professors and top scientists had been fired for criticizing evolution and that they had even been blackballed and thus couldn't get another job elsewhere. Yes, people have been fired, Wise warned. The issue is giving students and teachers the freedom to talk about arguments for and against evolution, he said.[62]

Senators on the opposing side argued that the bill was unconstitutional, that the new science standards already required a full discussion in all science classes, and that the Senate had no business interfering with the work of the experts who wrote the standards. They also thought the bill had religious roots. "This bill is not about evolution," Arthenia Joyner said. "It's not even really about academic freedom. What this bill is, is an attempt to bring the controversy of creationism versus evolution into our science classroom."[63]

Geller had quite a bit to say, starting with his reflecting on his high school days when he had acted in the play *Inherit the Wind*, which was based on the 1925 Scopes trial in Tennessee. "We were laughing at how backward those folks were," he said. "Fast-forward to 2008 and we're standing here on the floor of the Florida Senate in a year of unprecedented budget crises where we're cutting everything. And what are we debating on the floor of the Florida Senate? Evolution. I wonder what play is going to be written about this that we will laugh at decades from now." Storms took offense, saying that Geller's story about laughing at people he found to be "backward" was an illustration of what teachers and students now faced. "This is the person who advocates for tolerance on other occasions except in this occasion where people have different views."[64]

Scientific information seemed to be loosely defined in the bill, Geller said, and this could allow nearly anybody with a so-called fact to offer his or her views in the classroom. There is also the problem of state government trying to lure scientific research institutes to Florida at the same time senators were debating the validity of evolution. "They have to be questioning Florida's commitment to science," he said. Since the bill states that students shouldn't be penalized for their personal views, Geller wondered how far that could be taken. His interpretation was that students who encountered a test question about evolution could answer however they liked and not be marked wrong. "I don't think you can say 'this is my view of science' and expect to get full credit, although this bill will mandate that," he said.[65] And what happens after high school? What answers will those students give on college entrance exams?

Storms didn't respond to most of Geller's arguments in her closing statements. She concentrated on showing that there actually were Florida teachers affected by the controversy. She held up one e-mail from a national-board-certified science teacher who wrote: "To say that I have problems with evolution theory would be career suicide for me." The teacher wished to remain anonymous, but a second teacher had written to Storms and didn't mind being named. Wayne Gerber, who had taught science in Pinellas County for twenty years, filled three pages with criticisms of his school's biology textbook. "This is not accurate science," he concluded. "It is wishful thinking in support of a preconceived idea." With examples in hand, Storms declared that her bill was strictly about freedom of speech. "Do not be dissuaded," she said. "Do not be deceived. This is not about establishment of religion. The bill specifically says that it is not."[66]

The bill passed on a narrow 21–17 vote. One Democrat voted yes, while five Republicans voted no. It was now time to see what would happen to Hays's bill in the House.

"Back to the Dark Ages"

The House's second reading of Hays's bill kicked off at noon on April 25. Back when Hays's bill was debated in committee, Kiar was his primary opponent. On the House floor it was no different. Kiar offered two amendments for his fellow representatives to consider. First, he once again proposed that the words "critical analysis" be stricken. The same justification he used in committee, that the words make the bill unconstitutional, were presented again. When debate on Kiar's amendment opened, Keith Fitzgerald offered a spirited defense for it. Over time, he explained, science has developed specific and unique standards for critical analysis. For instance, the standards for critically analyzing poetry wouldn't apply in science. This is a concern, because the critical analysis called for by Hays is vague and invites religion into the science classroom discussion. "It would open up a can of worms that could politicize our schools in every direction." Gelber agreed, saying that the purpose of the bill was to "find ways to get faith taught in science class." Additionally, the board of education had created a set of science standards that clearly incorporate critical analysis already. "To mix science with faith does damage to both," he concluded.[67]

A few other legislators supported Kiar, pointing out that evolution is a valid, supported scientific concept and is integral to important research currently being done in Florida. "If we want to be cutting edge and attract bio-tech and high-level scientific jobs to the state of Florida we cannot be making a U-turn and riding our bicycles back to the dark ages where we can't talk about science," said Shelley Vana.[68]

Lawmakers against Kiar's amendment protested their opponents' claim that the bill was about religion. Instead, they insisted, it's about ensuring that children can think for themselves. John Legg blasted Kiar's amendment for sending the shameful message that students shouldn't be allowed to criticize and evaluate things, but rather passively accept what they're told in the classroom as gospel. Marti Coley said that she had been asked why gravity wasn't getting the same critical analysis attention that evolution was and she confidently stated that gravity was "pretty solid" and "proven beyond a shadow of a doubt." But when there are gaps in an idea, it warrants a closer look. "If we limit that critical analysis, if we are afraid to look at both strengths and weaknesses of a THEORY, how will we ever find any more information? It will simply be closed," she said.[69]

Hays defended his bill's use of the phrase "critical analysis." First, he shot down the "false assumption" that daring to question evolution equated to religion. The bill doesn't open the door to teaching religion, he said, and everyone who thinks it does is "barking up the wrong tree." Furthermore, removing "critical analysis" from his bill "guts the whole purpose of it." He then asked all scientists—or those "who think you are"—why they were afraid to have their theory analyzed. Kiar stuck to his guns, though. "The words 'critical analysis' is what makes this amendment unconstitutional," he said, "because it will permit and require teachers to teach opposite religious views of evolution, such as creationism and other religious doctrine. It will also allow atheist teachers to impose their nonreligious beliefs on our children . . . and I don't think there is anyone in this chamber that wants that to happen."[70]

A voice vote then shot down Kiar's amendment, but he had a backup ready. His second amendment asked that the word "scientific" be inserted before "critical analysis" in an effort to head off potential religious instruction in science classes. Hays called it a good amendment and stated that he would accept it without debate.[71]

Finally, debate on Hays's single-sentence bill began. A dozen lawmakers took turns giving their opinions, with more supporting the bill than were against it. A few tried to take a moderate position, such as Juan-Carlos Plana, who said the debate so far had featured arguments from both extremes with no attempt to find a middle ground. "I believe in the theory of evolution, yet I believe that God is responsible for it," he said.[72]

Gelber headed up the opposition by expressing his frustration that the legislature was "micromanaging schools" after the board of education and several subject-matter experts had already been entrusted to address the issue. Vana reemphasized her point that the bill would make it difficult for Florida to attract the "best and brightest." Fitzgerald was adamant that the bill's underlying claim that there is a scientific controversy that needed to be exposed was blatantly false. "This bill purports to correct a problem in the education of evolution that in fact does not exist." He then gave a short yet comprehensive and accurate lesson to the legislators on how science in general works, the scientific definition of the word "theory" as opposed to the colloquial usage, and evolution's solid standing in the scientific community. "And we all know what it's really all about. There are some folks, for religious reasons, who are threatened by the scientific teaching of scientific doctrine."[73]

Those in support of the bill sought to refute yet again the religious angle and emphasize the importance of critical analysis in schools. Thad Altman flipped the argument on its head, reasoning that when evolution is unassailable in the classroom, "what we are teaching is a religion in our schools." Will Weatherford picked up on Hays's suggestion that opponents were motivated by fear, a theme that became more dominant as the debate progressed. "If the evidence is so overwhelming, if it's so obvious, if the scientists are all pointing in the same direction that the theory of evolution is true, why are we scared to scientifically and critically analyze it in the classroom?" Kurt Kelly passionately argued that the bill clearly addressed science only and not religion. Opponents were just making "wild assertions," he said. "If we wanted to teach a religious perspective, we would teach from Psalms that says 'When I gaze into the night skies and see the works of his fingers, the moon and stars suspended in space, oh what is man that thou are mindful of him.' Yeah, we would teach that, but we're not."[74] Instead, the bill is asking for a scientific critical analysis, he said.

In Hays's closing arguments, he singled out Fitzgerald as the "personification of just why this bill is needed." Hays called his fellow representative's stance dogmatic as well as untrue. He suggested that everyone should see *Expelled* and said, "If you think there's not dispute in the scientific community you are badly fooled. I've got some ocean-front property in Phoenix I'd like to sell you.

"Dogma is saying evolution has no holes in it," Hays continued. He held up a letter he got from a member of one of the science standards committees stating that the bill was needed. He then repeated his assertion that evolution proponents were afraid of scrutiny, especially since there are so many problems with the theory. "I would also invite you to find anywhere in any documentation or anywhere in the earth where a fly has ever become a monkey, or where a monkey has ever become a person."[75]

The second reading of Hays's bill was just about to wrap up; however, it turned out that the debate wasn't quite over. There had been a procedural mistake early in the debate that had cut off a question-and-answer session too early. So, Gelber was permitted to quiz Hays for a couple of minutes. Gelber said that it wasn't clear to him what the bill was asking teachers to do. Does it require anything to be taught that isn't already outlined in the new science standards? Hays responded by reading excerpts from his bill, much like Storms had done in the Senate. Gelber then asked if the bill would allow the teaching of intelligent design in the classroom. Hays answered with a simple "No, sir."[76]

Gelber wanted a clearer picture of the bill's purpose, though. He reworded his initial question to ask what "other point of view" Hays was advocating so hard for that wasn't in the standards. Hays replied that he simply wanted to allow teachers to lead their classes in a critical analysis of the theory of evolution. The bill's purpose was to protect teachers' freedom of speech from a "bunch of radical people" and people like Fitzgerald who claim that "evolution is nothing but pure fact." Evolution was a far cry from "pure fact," Hays asserted. "I would submit to you that Charles Darwin did an excellent job of cataloguing all the species. A good librarian could do the same thing." Unsatisfied by Hays's slippery answers, Gelber repeated his point that Hays had been clamoring for "other views" without explaining what they were. It's fine that intelligent design is not one such view, he said, but is there some other view that the bill would allow that was not in the new science standards? No, said Hays, and that concluded the questioning.[77]

"What Does This Bill Do?"

Three days later the bill was before the House again for its third reading and final vote. The procedure calls for the bill's sponsor to first give a short explanation for the proposed bill. That is followed by a question-and-answer session during which lawmakers are permitted to ask the sponsor for information and clarification. They don't get into a back-and-forth debate during the questioning, but legislators have a tendency to phrase their questions in pointed and sometimes confrontational ways. Then there is a debate session that opens the floor to legislators to give speeches both long and short; however, there is no direct back-and-forth among them, despite the fact that the session is called a "debate." Next, the bill sponsor is allowed to give a closing speech. Finally, a recorded vote is taken.

Hays had faced tough questioning on his bill before, but this time he was grilled for a full forty-five minutes during the question-and-answer session alone. Gelber was the first to attempt to determine Hays's motives for introducing the bill. He couldn't understand what the bill actually did for teachers. Gelber asked what other "scientific theory" the bill allows teachers to present that they couldn't already teach without the bill, and Hays had no examples. Once again, Hays stated that teaching intelligent design wasn't allowed, that no religious concepts were allowed, and that the science standards wouldn't be affected. This left Gelber mystified. "What does this bill do?" he asked. "It allows the teachers to—without fear—expose the holes that exist in the scientific theory of evolution," Hays said.[78]

But Gelber wanted more. Rather than asking if intelligent design could be taught, as had been asked so many times before, he asked why it couldn't be taught. Is it because intelligent design isn't a scientific theory? he asked. This finally elicited a response with some meaning to it. "I wouldn't purport to say it's not a scientific theory, but, unfortunately, today in this country too many people are afraid of teaching religion in the classroom," Hays said. "I don't want religion taught to my daughters or my grandchildren in the public classroom, either. Religion should be taught in the home, should be taught in churches, synagogues and various other religious places. But, unfortunately, too many people—thanks to the Supreme Court's distortion of the First Amendment—too many people are afraid to even mention the fact or the possibility—the theory if you will—of intelligent design."[79]

Finally satisfied, Gelber made a motion to have their dialogue "spread upon the journal." This meant that he wanted a transcript of the exchange printed in the *House Journal* so that anyone in the future—judges and lawyers, for example—could use it to help discern the bill's legislative intent.

Audrey Gibson wanted to know if there had been some incident she wasn't aware of that had sparked the need for teacher protection. Hays chose not to go down that road. Instead, he focused on the future. "It's a preventive measure," he said. "You and I both know it's a whole lot easier to correct a problem before it occurs than it is to wait until after it's happened and the damage has been done."[80]

Then Hays faced nearly ten minutes of questions from Vana. Her main worry was that instead of protecting teachers, the bill would subject them to much more risk. Vana was a veteran science teacher, and based on her experience in the classroom she felt a clear definition of "scientific critical analysis" was needed. Since the bill would make such analysis mandatory, she didn't like how ambiguous it was. "How would I or any science teacher know exactly what to say to make sure they were complying with the law?" Hays countered that teachers are professionals who should be able to develop their own lesson plan and analysis "of whatever theory you are presenting." It would be an insult to teachers for the legislature to write out critical analysis instructions for them, he said.[81]

Vana tried to pin Hays down, though. Would she be protected "if I teach something that the scientific community does not agree should be taught" or if she included "with impunity" any non-mainstream, non-scientific theory? "This bill doesn't change your ability to bring in those theories any more than you can today," Hays said.[82]

Both became visibly frustrated as the questioning wore on. Hays repeated a few times that it was up to the teachers to determine what a satisfactory critical analysis is. "I don't know how to say it any other way, ma'am." But Vana kept probing in an attempt to see how far the interpretation of critical analysis could stretch. Opening the floodgates all the way, she asked if the bill made it permissible for teachers to include anything at all they personally felt like teaching in an evolution lesson. "That's something you take up with your school administrators," Hays said. And then after a three-second pause he added, "And the courts as well."[83]

That response set Vana off. Could she wind up in court if she taught something that others don't consider critical analysis or if someone thought her evolution lesson wasn't critical enough? Hays refused to outline the boundaries Vana sought, saying that he was a dentist, not a teacher. "I'm not going to get into micromanaging the educational system."[84] Hays insisted that his bill wouldn't land more teachers in court more often, and Vana was finally forced to give up.

The next few lawmakers took just a minute or two each to dig into the bill's underlying intent. Tony Sasso conjectured that this was all about wanting to find holes in evolution, which would naturally lead to a discussion of religion in the science classroom. Obviously tired of this accusation, Hays told him he was wrong and defended the simple and clear language in the bill. "I don't mean to sound like a smart aleck, but I would refer you to an English teacher to help you understand the meaning," Hays said.[85]

When it was Fitzgerald's turn, he took his time building a case. He confirmed with Hays that there had been no previous documented cases of teachers facing repercussion and that it was up to the teachers to define critical analysis. With that foundation established, he then asked: "You just told us you have a bill here that corrects a problem that doesn't exist in a way that will have no enforcement mechanism whatsoever. So what is the point of passing this bill?"[86] It empowers teachers to teach a scientific critical analysis, Hays explained.

Fitzgerald followed up, asking how Hays knew that teachers weren't already empowered to do so. What, exactly, is the evidence that a problem even exists? "I would say the evidence is subjective," Hays replied. He also explained that the massive debate and "all the alarm" the issue had sparked among Florida citizens and legislators demonstrated that there was a need. Fitzgerald wasn't buying it, though, and asked if teachers or their unions support the bill. "I have no idea. I haven't asked them," Hays replied.[87] But he then suggested that Fitzgerald watch *Expelled* for examples of people being fired for questioning evolution.

Fitzgerald went on to quiz Hays about whether his bill permitted the exposure of holes in the theories of quantum mechanics in physics, plate tectonics in geology, and supply and demand in economics. Hays said no each time. Then why—if religion is not a motivation—is evolution being singled out from the hundreds and thousands of other possible theories that have problems that can be questioned, Fitzgerald asked. "Because of the same answer I gave you a while ago," Hays said. "So many people are afraid that this questioning the theory of evolution is going to require or bring into the classroom the teaching of religion. And that is not true."[88]

Hays then faced hostile questions from a fellow Republican, which was rare in debates about his bill. Carl Domino had heard Hays recommend *Expelled* to lawmakers several times now, and Domino had taken that advice. Hays was visibly uncomfortable as Domino cornered him with pointed questions about the movie. Domino asked if the movie showed any high school teachers being fired. "I don't remember that much detail about it," Hays said. Wasn't the movie more about people trying to obtain grant money to study alternatives to evolution, Domino asked. "Quite honestly," Hays replied, "I don't think that today is the appropriate time to really discuss the movie. But in my particular judgment, the message in that movie was: Look out folks! The freedom of expression, the freedom of speech, is something you really need to jealously guard, and that's what this bill does."[89]

Later, Hays's frustration showed even more as he sparred with Elaine Schwartz, who was a lawyer. She asked, "Are you aware that this bill will probably be a subject of a great deal of litigation involving our schools?" His testy response was: "I would think your profession would welcome that." But after further questioning he stated, "The purpose of this legislation is to minimize the likelihood of lawsuits."[90]

In all, fifteen legislators questioned Hays, with the majority of them opposed to the bill. Once the questioning petered out, the floor was opened for debate speeches.

"Meatball Surgery"

Fitzgerald was the first to speak, and he came out swinging. "The sponsor of this bill told us the other day that there was no evidence of evolution turning a fly into a monkey. But this bill shows definitively that bad bills can turn legislators into monkeys. This is silly legislation." That earned him a scolding from speaker pro tempore Marty Bowen, who was presiding over the debate. Fitzgerald was warned to abstain from name-calling.[91]

Allowing teachers to teach whatever they want "is to promote incompetence," Fitzgerald continued. Teachers would not be held responsible if they taught that wishing hard could make you fly or that two plus two doesn't equal four. "To suggest that a teacher has the freedom to bring ideas in from other fields, choose not to teach what science actually teaches, is a very dangerous concept." He implored everyone to vote against "this embarrassing bill" that did nothing to solve a problem that didn't exist. Taking up this theme, Domino pointed out that many people believed that the Holocaust never happened or that the moon landings were faked. "There are a lot of strange things out there that I don't want my teachers teaching," he said.[92]

A few others didn't like what they viewed as the bill's meddlesome nature and the many questions that still clouded it. "I think this bill ultimately is playing meatball surgery with our curriculum," Gelber said. Audrey Gibson felt the bill would open a Pandora's box. "The sponsor seems not even to know what the definition of critical analysis is. Well, if you can't define a thing then how in the world can you legislate it?" Dorothy Bendross-Mindingall, a former teacher and school principal, was worried that the bill would create "havoc and mayhem."[93] Additionally, the bill's call for critical analysis would require appropriate teacher training, which the bill makes no provision for, she explained.

The debate leaned heavily toward those in opposition, with thirteen lawmakers urging a no vote. On the other hand, six speakers supported the bill, with most of them arguing for keeping scientific inquiry open. Frank Attkisson said that evolution is a theory. "A theory is a question. It is not fact. If it is fact, we call it fact. And everybody today is trying to say we should present theory as fact," he said. He claimed that Charles Darwin, who advocated for examining "facts on both sides of each question," would thank Hays for sponsoring this bill.[94]

Defending Hays's assertion that the bill was a preventative measure, Paige Kreegel compared it to vehicle safety. Doesn't seatbelt use prevent something that hasn't happened? He also said that although he believed in evolution, critical analysis was still needed, since the theory of evolution itself was evolving. Doug Holder claimed that the hostility toward the bill was strictly partisan. Opponents would stand up for a discriminated liberal cause, but they don't like Hays's bill because it's based in a conservative issue, he said. Coley assured everyone that there was no hidden agenda and that evolution could be analyzed without mentioning religion. "I'm not sure what we're so afraid of," she said.[95] Altman agreed, saying that a vote for the bill was a vote for academic freedom and free speech.

The heated debate was finally over, and the floor was turned over to Hays to give a closing statement. Although there was little doubt that it would pass, Hays used the opportunity to punctuate the lengthy conflict with a clear view of his governing principles. "The founders of our country were God-fearing men who believed that no government should be afraid of religion," he began. "No government should establish a religion, and they should not work against religion. Most of the people who have spoken today appear to be against religion." Loud groans of protest interrupted Hays, forcing him to tell everyone to "wait just a minute" and allow him to finish.

"I've said to you I'm not in favor of bringing religion into the public classrooms, also. But I even think that my belief on that is an aberration of what the founding fathers of this country wanted. . . . My question to you today is: what are you afraid of? Are you afraid that our students are going to learn how to critically analyze a theory? That's what you seem to be saying. And I suggest to you that Charles Darwin was definitely a scientist. And a true scientist will put his theory out there and invite you to pick it apart and say how many different ways can you disprove that. A true scientist is searching for the truth. And that's what we should be encouraging our students and our teachers in the public schools in this state to search for every day: search for the truth."[96]

With that, the vote was taken and the bill passed 71–43. The Senate's approved bill had been sent to the House, but the House refused to even consider it. Instead, the House now shipped its approved, single-sentence version to the Senate.

"A Lot of Explaining to Do"

However, it was Monday, and the legislative session was set to close on Friday. Both chambers had previously rejected the other's version of the bill, and neither Hays nor Storms seemed confident the dueling bills could be reconciled. Storms said that she knew the Senate wouldn't accept the House's version, so all she could do was ask the House to reconsider. "I have no choice," she said. "And I believe in miracles."[97]

The House bill was "simpler and more straightforward," stated a memo issued by the House Majority Office. "Also, unlike the Senate bill, this bill does not create any new 'rights' for teachers."[98] However, senators were uncomfortable with how the House version went beyond just allowing challenges to evolution and actually required critical analysis in the schools. On Thursday the Senate officially refused to consider the House's bill language and bounced it right back.

With time running short, Storms was mystified as to why the two chambers had diverged in the first place. "At the beginning of the session, both the House and Senate agreed to travel by caravan to an important destination by way of a difficult and rocky terrain," she told the *Florida Baptist Witness*. "Suddenly, as we drew near our destination after a tumultuous trip, the House vehicle veered off of the sure path to our destination and merrily called out, 'We know a better way! Follow us!' However, the Senate vehicle is filled with senators who want to stick to the known path since the destination is clearly in sight. So, the Senate vehicle presses on, arriving at the agreed upon destination and hoping that the House makes it there in time."[99]

As John West, senior fellow at the Discovery Institute, watched the bills falter, he wondered if something shady was happening behind the scenes. "I'd say someone in the House, in the name of trying to push this bill, is actually trying to kill it," he said. "If nothing passes, I think there will be a lot of explaining to do."[100]

When the end-of-session ceremony was conducted on Friday, any chance of a new law challenging the teaching of evolution died. The House hadn't responded to the Senate. Florida Citizens for Science noted on its blog that day: "Let us take a moment of silence for House Bill 1483 and Senate Bill 2692, the deceptively named 'academic freedom' bills. Time of death: 6 p.m. I doubt they will rest in peace, though." The bills' supporters had little to say after the last-minute failure. The *Florida Baptist Witness* felt that the effort might have been destined for failure from the start. "Starting a major piece of legislation only two weeks before the session began made passage difficult."[101]

A question remained: Would "academic freedom" bills return next year?

11

"Standing Up for the Little Guy"

Darwin Day is an unofficial holiday in mid-February that celebrates the famous naturalist's birth, and many museums, universities, and clubs worldwide host celebrations. There were a few events in Florida in 2008 that coincidentally served as ironic counterpoints to the evolution battles raging in the state capital at the same time. For instance, there was a day of guest lectures at Florida State University and birthday cake at Fern Forest Nature Preserve in Coconut Creek.[1]

An event hosted by the Friends of Brooker Creek Preserve in Pinellas County was to feature University of South Florida professor Dr. Lorena Madrigal. The anthropologist was obviously expected to talk about Darwin and evolution, which were subjects she had extensive professional knowledge about. However, a week before the celebration she was disinvited. "They told me very clearly they felt their budget was in danger if the lecture took place," Madrigal told the *Tampa Tribune*. The reporter verified that with Pinellas County's director of environmental services, William Davis. "Her topic was about evolution," he said. "I flinched on that. I canceled her out after discussing it with my supervisors. We are not the platform for debate on creationism versus evolution."[2]

This was Madrigal's second brush with the controversy between evolution and creationism. She had participated in a debate event about the subject in 1992, which she later regretted. Now she found herself in the middle of a controversy she hadn't foreseen. "I got invited a long time before the event, and at the time there were no hints that anything was wrong at all," she said. "I don't think they ever thought that such problems would arise." Madrigal was notified of cancellation by a staff person. "He apologized numerous times, and asked if I would consider coming back if the storm passed," she recalled. "I said yes. Public service is part of my assignment as a professor."[3] The storm did pass, and Madrigal gave a Darwin Day talk in 2009, the bicentennial of Darwin's birth and sesquicentennial of the publication of *On the Origin of Species*.

Despite the kerfuffle over evolution in the opening months of 2008, there was little excitement for the subject throughout the remainder of the year. A few candidates for local school boards faced questions on the subject, but regardless of which side they stood on, they didn't turn it into a major campaign issue. In October, the ACCENT Speaker's Bureau at the University of Florida hosted a debate titled "Science, Man and God: A Creation vs. Evolution Debate." Dr. Douglas Jacoby, a Christian author and minister who was billed as a creationism advocate, faced off against Dr. Michael Shermer, founding publisher of *Skeptic* magazine and an evolution advocate.

An overflow crowd of more than a thousand people attended, but the debate had a slight twist to it. Rather than take a hard-line creationist approach, Jacoby argued that science and religion are compatible. He called himself an "evolutionary creationist" who believes that God used the process of evolution in his creation. Shermer countered that evolution is a purely natural process with nothing supernatural involved.[4]

"Which Political Party Has Evolved"

Early in 2009 it looked like it would be another banner year for the evolution conflict in Florida. Sen. Stephen Wise announced in February that he planned to introduce a new bill about evolution during the next legislative session. The bill would require teachers to introduce intelligent design whenever evolution was discussed in the classroom. "If you're going to teach evolution, then you have to teach the other side so you can have critical thinking," he told the *Florida Times-Union*.[5]

Sen. Stephen R. Wise (R-Jackson-
ville) filed a bill in 2009 that would
require "a thorough presentation
and critical analysis of the scientific
theory of evolution" in Florida public
schools. (Courtesy Florida Senate,
official photo.)

The sixty-seven-year-old senator had been a fixture in the capital since winning a seat in the House during a 1988 special election. In 2001 he was elected to the Senate. Wise had co-sponsored Sen. Ronda Storms's 2008 "academic freedom" bill. In the light of that failed effort, his plan for 2009 was to make sure that intelligent design bills in both houses of the legislature stayed on the same course. Even though he hadn't been a vocal figure in the 2008 bills' efforts, Wise had received a lot of hate mail. "You'd think I'd never gone to school, that I was a Cro-Magnon man, that I just got out of a cave or something," he said.[6]

The *Florida Times-Union* then tracked down Rep. D. Alan Hays—who was the 2008 bill's sponsor in the House—and learned that he would support Wise's new effort in his chamber. Hays believed that this time the bill would be successful. "The thing we learned last year is that, number one, we must keep the discussion scientific," he said. According to the newspaper, Hays was primed for the push, having recently been a guest speaker at an "Intelligent Design and Public School Curriculum" symposium at Liberty University School of Law in Virginia.[7]

The reactions to Wise's announcement were predictable. Larry Spalding from the ACLU said his group would certainly challenge the bill. "In tough economic times, why would you pass a law that would invite a very expensive lawsuit?" Rep. Rick Kriseman, a Democrat, wished that Florida Republicans in the legislature would join Democrats "in the 21st century." If the bill was filed, it "will give voters a chance to see which political party has evolved," he said.[8]

Wise defended his proposal during a radio interview a week before the legislative session started. "I always like the story, the person says, well, 'You know, we came from monkeys, we came from apes.' Well, why do we still have apes if we came from them? And those are the kind of questions kids need to ask themselves." Wise had heard about college students failing classes where evolution was presented because they didn't believe in it. "And, you know, there's more than one theory on this thing. And the theory is evolution, the other one is intelligent design."[9]

However, when Senate Bill 2396 was finally filed there was no mention of intelligent design. Instead, Wise used the same tactic Hays had the previous year: require critical analysis. The bill would amend the Florida statute outlining "required instruction" in the public schools by adding a line mandating "a thorough presentation and critical analysis of the scientific theory of evolution." This was similar to what Hays had done when he had completely revised his bill in 2008.

This time there was a new twist, though. Wise included in his bill two other amendments to the statute. One would require to be taught "the historical context in which the Declaration of Independence was drafted and signed." The other would require that "conflict-resolution management" be incorporated into elementary school character-development programs. So, rather than giving the bill a title directly relating to critical analysis or evolution, it was called "Relating to Educational Instruction" because it encompassed three separate subjects.[10]

Two organizations that were now veterans in the war over the teaching of evolution spoke out against the new bill. Florida Citizens for Science fired off a news release in response to the filing: "Wise's antievolution bill is an insult to citizens who are tired of stomping over the same ground over and over again. The Florida Board of Education and last year's state legislature have already debated the teaching of evolution ad nauseam. To insist on bringing this up again is irresponsible because it will distract our lawmakers from the important tasks at hand, and could burden one of our school districts with a million dollar legal bill." Joining them was the Florida Academy of Sciences, whose president, Richard L. Turner, released a position statement opposing the bill: "SB 2396, in effect, leaves the door open for the introduction in the public school curriculum of nonscientific and covertly religious doctrines. The proposed bill would be damaging to the quality of science education of Florida's children and the scientific literacy of our citizens."[11]

The bill was sent to two committees: Education Pre-K–12 and Education Pre-K–12 Appropriations. As the legislative session progressed, though, no action was taken on the bill in either committee, and no companion bill appeared in the House. When the *Tampa Tribune* asked Hays why he didn't revisit the issue, he said that there were more important things to take care of and that there was a limit on the number of bills that House legislators could introduce. By the end of the session the Senate bill had died in committee.[12]

"A Refreshing Renaissance"

As the evolution controversy faded on the state level stage, it still simmered at the local level. In 2006, Dr. Thomas Woodward of Trinity College had organized a "Darwin or Design" conference that promoted intelligent design, and in 2009 he unveiled a fossil collection on permanent display at the Christian school's campus library. "I'm seeing here a refreshing renaissance in the Christian community of wanting to learn more about science, and seeing it as a friend and not a foe," Woodward said. The collection features fossilized tree sap with embedded insects, various dinosaur bones, and specimens such as trilobites dating back to the Cambrian period. Woodward said he was amazed by the new discoveries coming from studies of the Cambrian, when life's diversity really started to flourish and leave traces behind. "The more we check it out, the more diversity we find. That's data, that's information. That's scientific discovery, and I think it's exciting because it tends to support the idea of a creator."[13]

Florida Citizens for Science stayed active in 2009, too. The organization joined with the National Center for Science Education, a national pro-evolution organization based in California, and the Coalition for Science Literacy, headquartered at the University of South Florida, to run a four-day teacher workshop in Hillsborough County. Called "Controversial Issues in the Science Classroom," the workshop had a strong focus on problems teachers may encounter while covering evolution, but it also touched on global warming and how these controversial subjects can spill over into other science disciplines, such as physics, geology, and astronomy.[14]

The founder of the Coalition for Science Literacy, Dr. Gerry Meisels, was interested in learning how Florida's new science standards, which explicitly included evolution, were being received at the classroom level. In 2010 he co-published a study titled "Florida Teachers' Attitudes about Teaching Evolution," which examined the results of 353 surveys completed by teachers from the state's elementary, middle, and high schools. The study found that "only 72 percent of the teachers agreed that evolution is a central organizing principle of biology."

The science standards required that elementary schools introduce evolution concepts, and Meisels's study revealed that many Florida teachers at that level weren't comfortable with the material. Only 69 percent of elementary teachers felt they understood evolution well enough to teach it, compared with 88 percent in middle schools and 95 percent in high schools. Further complicating the issue was the finding that only 61 percent of elementary school teachers agreed that the Earth is at least 4 billion years old. The numbers were much higher in middle schools, 85 percent, and high schools, 95 percent. This led Meisels to conclude: "Now that evolution has become a Big Idea in Florida's science standards beginning at the elementary level, helping these teachers become more comfortable with and knowledgeable about evolution is increasingly important."[15]

"Just Plain Bad Biology"

Now that the new science standards were being eased into classrooms, schools needed updated textbooks to match the standards. Instructional materials advisory committees pored over a wide variety of science books and supplementary products in 2010 to compile a list to be

presented to Education Commissioner Eric J. Smith. He would then review and approve for adoption a list of materials that school districts
would choose from when buying books for their classrooms.

The vast majority of the books committee members examined were
of high quality and addressed evolution thoroughly, in keeping with the
new standards. But one book raised a red flag: *Life on an Ocean Planet*,
a marine sciences book published by Current Publishing in California.
In it was a two-page informational sidebar titled "Questions about the
Origin and Development of Life." Some committee members pointed
out that the sidebar was riddled with factual errors and contained arguments commonly associated with creationism. For instance: "Skeptics
observe that general evolution doesn't adequately explain how a complex structure, such as the eye, could come to exist through infrequent
random mutations." Jonathan Smith, a Florida Citizens for Science
board member, said, "This is a standard creationist trope, well known to
be wrong."[16]

The committee approved the book on a 7–2 vote, but with the condition that the two questionable pages be removed. One of those who
voted no regardless of conditions was Alachua County School Board
member and retired teacher Eileen Roy, who stressed that if there were
"egregious" errors in just those two pages, the rest of the book was sus-
pect as well. "I was there to make sure we had textbooks that are scientifically accurate," she said. The other no vote came from Clay County
teacher David Campbell, who was also a Florida Citizens for Science
board member. "The first thing I noticed is just plain bad biology," he
said.[17]

Florida Citizens for Science quickly took action. President Joe Wolf
sent a letter to Commissioner Smith asking him to carefully review the
book before approving it. "A textbook's job is to present the current
state of science so that students can engage with contemporary science," Wolf wrote. "However, this textbook's treatment of 'Origin and
Development of Life' is clearly bad science and bad pedagogy. The sidebar is simultaneously actively misinforming, at odds with state standards, and ultimately irrelevant to marine science. It should be removed
entirely, as there is so little information that is either correct or useful
to make it worth retaining."[18]

Dean Allen, the vice president of Current Publishing, responded to the criticism, saying that there had been no intention to undermine evolution. The sidebar was merely meant to be a "critical thinking exercise for students." However, once the errors were pointed out, Allen agreed to remove the pages. "That's the only two pages that went off track a little bit," he said.[19]

"A Theory of Whatever"

When the 2011 legislative session kicked off, a familiar bill reappeared. Wise filed Senate Bill 1854, which would mandate "a thorough presentation and critical analysis of the scientific theory of evolution," just like his attempt in 2009. The language on the Declaration of Independence and character development was there, too.[20] However, there were a few new circumstances that could affect the bill's fate. Wise was now the chairman of the Education Pre-K–12 Committee, one of two committees the bill would need to pass through. The Senate Budget Committee was the other, and Wise was a member there along with Hays, who had recently been elected to the Senate. Wise also felt that the legislative environment was better for the bill this year, since there were even more conservative lawmakers serving than before.

In 2009, Wise clearly thought intelligent design was a solid alternative to evolution, even though his bill then didn't directly address it. In 2011 he was vague in describing what could be taught while denying that religion was in any way involved. He told the *Tampa Tribune*, "We're not saying you ought to be a Muslim, you ought to be Jewish, you ought to be Christian or you ought to be Baptist or Episcopalian; what we're saying is here's a theory, a theory of evolution, a theory of whatever, and you decide." Wise also referred to evolution alternatives as simply "nonevolution."[21]

Florida Citizens for Science and the Florida Academy of Sciences publicly opposed the bill, restating most of the same arguments they used two years prior. A news release from Florida Citizens for Science emphasized Wise's "theory of whatever" and "nonevolution" statements: "Sen. Wise needs to be honest and forthright. What kind of 'critical analysis' is he really wanting?" Howard Simon, executive director of the Florida

ACLU, said the bill would not go unchallenged if it became law. "The mischief that this legislation does is that it tries to entice local county school boards into violating the constitution," he said.[22]

Despite Wise's early optimism, though, the bill died out just like the 2009 version had, without even making it onto a committee calendar.

Split Personality

Nine decades after the 1923 Florida legislature first tackled the problem of evolution in the state's schools, the battle continues. In late 2011 an "Answers in Genesis" conference at Hibernia Baptist Church in Fleming Island attracted nine hundred attendees who wanted to learn about creation, evolution, and dinosaurs from a strictly biblical viewpoint. Speaker Mike Riddle lectured that dinosaurs and humans lived together and that dinosaurs could have been on Noah's ark. "You rule out the truth if you cannot accept the truth," he told his audience. "You accept anything in its place. That's what evolution is." When asked why public schools "teach false information," Riddle replied: "The government, basically, is at war with Christianity."[23]

Those who had fought hard against evolution in 2008 sought elected positions in subsequent years. Terry Kemple ran for a seat on the Hillsborough County School Board in 2010 but came in last out of four candidates, receiving only 19 percent of the vote. He tried again in 2012, this time placing second out of six candidates and earning a runoff election against a longtime incumbent. Even though Kemple made strong conservative principles the bedrock of his campaign, the teaching of evolution wasn't one of his targets. In response to a reporter's question he did say that "the fact that the state requires us to teach that evolution is the be-all and end-all is a travesty." However, he admitted that the subject couldn't be effectively resolved on the local level.[24] Kemple lost the runoff, but he did round up a respectable percentage this time, 43 to his opponent's 57.

Activist parent Kim Kendall ran for a House seat in 2012. It turned out that both of her opponents in the Republican primary had similar views on evolution. Kendall predictably responded to a question on the subject by saying: "Evolution standards are being thrust on our students in science class. Academic freedom allows our students to question that science has it right." One of her opponents, Ronald Renuart, agreed: "Evolution is still a theory. It should be taught as a theory, not as a fact. Creationism, divine intervention—a lot of people share this belief." The other candidate, Mike Davis, said, "The school should present both ideas."[25] The vote was close, but Renaurt won and Kendall finished last. Renaurt went on to win in the general election and earn a seat in the House.

The Florida Department of Education got a shock in 2012 when the Fordham Foundation released its latest review of the states' science standards. This was the first time Florida's revised standards would be officially graded. The good news was that the standards no longer rated an F grade. The bad news was that the standards only bumped up one grade, to a D. Although the Fordham review team praised Florida's coverage of evolution in the standards, including human evolution, which few other states even mentioned, it found several other areas substandard. "Florida's standards evoke a split personality," the report summary stated. "The document starts out well at the primary level, but in the higher grades it weakens into poor organization, ambiguous statements, and basic errors. One has the impression that the writers were pushing the limits of their scientific expertise at the higher grades. Taken as a whole, the document does not provide a solid foundation for a rigorous K–12 science curriculum."[26]

Individuals from the Fordham Foundation had unofficially reviewed the standards in 2008 and had given them a lot of praise, leaving the impression then that the standards were much better than a D. This confused personnel at the department of education, including Mary Jane Tappen, who had taken a lead role in the revision process. Nonetheless, Tappen wanted to make sure the science standards were the best they could be. "Certainly, if we have any errors or content that leads to misconception, we need to make those corrections immediately," she told *Education Week*.[27] How that would be accomplished, though, was unknown.

The surprisingly low grade bothered board of education member Roberto Martinez so much that he called the Fordham Foundation to ask about it. After fielding multiple calls about the issue, the foundation revised Florida's grade. A representative from the foundation explained to the *Orlando Sentinel* that the error was on their end. "Florida had been 'double faulted' for the same problem with its chemistry and physics standards," she said. "Fixing that, gave the state one more point, which was enough to bump it up to a C."[28]

"A Blasphemous Fallacy"

"The Florida Columbia County School District has been dodging a public debate between the Scientific Theory of Evolution and the Holy Bible," said a December 2011 letter to the editor in the *Lake City Reporter*. "They are using our Florida taxes to thrust their false, blasphemous, anti-Scriptural teaching upon Columbia High School and Forth White High School Biology 1 students, who have to learn this evolutionary explanation of how they came into being." Thus did Lake City resident Kenny Merriken publicly announce his crusade against the teaching of evolution. It was just the opening salvo in a long-term campaign.

In January 2012, Merriken stood before the Columbia County School Board to protest the teaching of evolution in the local schools. His presentation didn't prompt action, because he only talked for a few moments during the public-comments portion of the meeting and his subject wasn't on the official meeting agenda. Also, his stance was blatantly religious. "The heart of the matter of the public debate on the Scientific theory of Evolution," Merriken said, "is that it is a blasphemous fallacy which tried to get around the fact that the King of the Jews created, in His image, all of the Columbia County School District students under your authority," he told the board.[29]

For several months Merriken picketed the school district offices on weekday mornings, during school board meetings, and sometimes on weekends. "I'm standing up for the little guy," he said. School officials were, of course, well aware of Merriken's actions, but the school district superintendent believed Merriken was wasting his time on the local level. "We use state-approved textbooks and we teach state-approved curriculum," Michael Millikin told the local newspaper. "We are simply doing what the state has charged us to do." Merriken wasn't deterred, even after a setback during a high school open house during which he was issued a no-trespass warning.[30]

Merriken spent thousands of dollars on radio and newspaper advertisements. The newspaper ads became a regular feature month after month. His ad in April 2012 implored fellow taxpayers to research what was being taught to their children: "With our taxes, Florida School Districts will be testing Biology 1 public school students commencing in 2012 concerning the blasphemous fallacy of the Scientific Theory of Evolution, which is contrary to the Word of God." He challenged anyone from the school district, including teachers, to "a public debate between The Scientific Theory of Evolution and the Holy Bible."[31]

A few months later he set his sights on the candidates for Columbia County school superintendent. His June 2012 ad asked the candidates, "Am I correct when I proclaim to you that Columbia High School students are created in the image of God and that none evolved from a hominid?" Merriken demanded a clear answer of yes or no. But he complained that no one would give him that answer, instead providing what he termed as "politically correct sidestep responses."[32]

Merriken became a fixture at school board meetings starting in December 2012, giving his opinion about evolution during the public comments time.[33] As of this writing, his appearances haven't prompted any official responses.

Epilogue

It's difficult to write an ending to a book when the tale it tells isn't finished. The last chapter left off with Kenny Merriken's crusade in Columbia County, but it's clear Merriken is just getting started. A few decades ago, Rev. Clarence Winslow set out on a path similar to Merriken's. Winslow stumbled a bit but eventually found his footing and then came within a whisker of succeeding. Even after that failure, Winslow kept on going until his death. Remarkably, he didn't get started on his lengthy crusade until he was about sixty-four. Merriken started at sixty-three, and it seems that he has Winslow's energy and drive. Will he keep marching on for decades, like Winslow? Will he eventually find a path to success? Only time will tell.

There are a few other pots simmering on the back burner. An organization called World Changers of Florida went to court in 2010 to secure permission to distribute Bibles on public school campuses in Collier County. The group won there and then added Orange County schools to their distribution campaign in 2013. In January 2013 the *Orlando Sentinel* reported that World Changers' president, Jerry Rutherford, might be looking for a new challenge: "Rutherford, who grew up and attended public schools in Orange County, said he'd like to see schools incorporate bible verses and religious teaching in schools as well as offering Creationism as a competing theory to evolution." This is confirmed by the organization's website, where their first objective is "To support the biblical account of Creation, including having Creation theory taught in our public schools."[1]

In chapter 8, I briefly explained three voucher programs in Florida and how tax money from them went to some private schools that teach various forms of creationism in science classes. The issue never seemed to raise many eyebrows. More recently, though, a young man from Louisiana, Zack Kopplin, began to draw attention to the potential problem in several states. He compiled a list of private schools that teach creationism while accepting tax-supported vouchers. Kopplin explained in January 2013, "So far, I have documented 310 schools, in nine states and the District of Columbia that are teaching creationism, and receiving tens of millions of dollars in public money through school voucher programs." More than half of these, 164, are in Florida.[2] His story was picked up in the national news and caused a bit of a stir. Surprisingly, the Florida media as of this writing have ignored Kopplin and his list.

These are just a few of the pots threatening to boil at any time. We also can't forget that the state science standards will be revised again in future years as part of a regular review cycle and that many of those who fought against evolution in the standards in 2008 are still around and active. There is also a chance that the state standards will be replaced by national science education standards, which have evolution prominent in appropriate sections. However, Florida education officials have been very guarded about their plans. Perhaps they're wary after the dramatic conflicts in 2008.

In other words, it's not a matter of whether more chapters will be written, but when. I started work on this book in 2008 after the battles detailed in chapters 9 and 10 took place. I was curious as to whether the conflict over the teaching of evolution had ever been an issue in Florida before the events of 2008. I was genuinely surprised to dig up all that is now documented in this book. I had no idea so much had happened! The constant flow of new discoveries made this topic a thoroughly enjoyable one to research and write. And over the years that it took to finish this project I've come to understand something important. This war over the teaching of evolution isn't an intellectual struggle but rather a deeply emotional one. It speaks directly to who we are as human beings, a society, a country, and individuals. Is there a divinely inspired purpose for our existence, or are we just accidents of physics and chemistry?

The opposing sides were absolutely confident in their answers to that question in the 1920s, and today's warriors are just as cemented in their views. Their tactics have changed, but their passions have not. It's a certainty that someone's passions will spark the next battle. My fingers are hovering over the keyboard, ready to chronicle it. I doubt I have long to wait.

Acknowledgments

Thank goodness for packrats! Nancy Marsh, retired Hillsborough County Schools supervisor of secondary science, still has boxes stuffed full of documents from back in the early 1980s when her school board required the teaching of scientific creationism alongside evolution. Without her personal recollections and her gracious willingness to share that magnificent treasure trove of files, parts of this book would have been incomplete.

The National Center for Science Education has an extensive library archive that I took advantage of and, even more importantly, a very knowledgeable staff. Deputy director Glenn Branch read an early version of the manuscript and gave me a three-page list of suggested corrections and clarifications. It took me several days to check everything off that list, but it was time well spent.

My fellow Florida Citizens for Science board members have been a wonderful support network as I spent nearly four years researching and writing.

Notes

Preface

Epigraph: Associated Press, "Bryan Stormy Petrel of Presbyterian Assembly," *St. Petersburg Evening Independent*, May 19, 1923.

1. Dr. C. Francis Byers, interview by Robert Johnson, August 2, 1973, 26, University of Florida Samuel Proctor Oral History Program, accessed October 26, 2012, http://ufdc.ufl.edu/UF00005904/00001.

2. Ibid.

Chapter 1. "Between the Devil and the Deep Blue Sea"

1. Jack Mills, "The Speaking of William Jennings Bryan in Florida, 1915–1925," *Southern Speech Journal* 14, no. 3 (1949): 139.

2. Williams Jennings Bryan and Mary Baird Bryan, *Memoirs of William Jennings Bryan* (Whitefish, Mont.: Kessinger, 2003), 479; and Associated Press, "Baptists Join in Arguments on Evolution," *St. Petersburg Times*, May 16, 1922.

3. "Miami Christian Council Has First Winter Session," *Miami Herald*, November 28, 1922; "Protecting Our Children," *Miami Herald*, November 30, 1922.

4. Associated Press, "Bryan Stormy Petrel of Presbyterian Assembly," *St. Petersburg Evening Independent*, May 19, 1923.

5. Florida House of Representatives, *Journal of the House of Representatives*, 1923, 482–83, http://www.myfloridahouse.gov/Sections/Search/HistoricalJournal/HistoricalJournal.aspx.

6. Edward J. Larson, *Summer for the Gods: The Scopes Trial and America's Continuing Debate over Science and Religion* (New York: Basic Books, 2006), 47.

7. Florida House of Representatives, *Journal*, 1923, 1840; and Mills, "Speaking of William Jennings Bryan," 151–52.

8. Michael Lienesch, *In the Beginning: Fundamentalism, the Scopes Trial, and the Making of the Antievolution Movement* (Chapel Hill: University of North Carolina Press, 2007), 128.

9. Bryan and Bryan, *Memoirs of William Jennings Bryan*, 482.

10. Florida House of Representatives, *Journal*, 1923, 1853–54.

11. "W. J. Bryan Protests against Ridicule," *Tampa Morning Tribune*, June 22, 1923.

12. Samuel Proctor, "William Jennings Bryan and the University of Florida," *Florida Historical Quarterly* 24 (July 1960–April 1961): 3, 4, 8.

13. William Jennings Bryan, *The Menace of Darwinism* (New York: Fleming H. Revell, 1921), 22–23.

14. Mills, "Speaking of William Jennings Bryan," 150.

15. Proctor, "William Jennings Bryan," 12.

16. Dr. C. Francis Byers, interview by Robert Johnson, August 2, 1973, 26, University of Florida Samuel Proctor Oral History Program, http://ufdc.ufl.edu/UF00005904/00001.

17. Dr. John Henry Davis, interview by Robert Johnson, September 4, 1973, 3, University of Florida Samuel Proctor Oral History Program, http://ufdc.ufl.edu/UF00005905/00001.

18. Florida House of Representatives, *Journal*, 1925, 1267.

19. Ibid., 1579.

20. Randy Moore, *Evolution in the Courtroom: A Reference Guide* (Santa Barbara, Calif.: ABC-Clio, 2001), 30.

21. Eric W. Gritsch, *Toxic Spirituality: Four Enduring Temptations of Christian Faith* (Minneapolis: Fortress Press, 2009), 55.

22. Benjamin C. Gruenberg, *Elementary Biology: An Introduction to the Science of Life* (Boston: Ginn and Company, 1919), 470.

23. Adam R. Shapiro, "Civic Biology and the Origin of the School Antievolution Movement," *Journal of the History of Biology* 41, no. 3 (2008): 422.

24. Ibid.

25. Ibid., 429, 415.

26. "Battle against Evolution Teaching Is On in Florida," *St. Petersburg Evening Independent*, December 29, 1925.

27. "Crusaders Plan Campaigns against Evolution Teaching," *St. Petersburg Evening Independent*, January 2, 1926.

28. Adam R. Shapiro, "The Evolution of the 'New Civic Biology,'" Committee on Conceptual and Historical Studies of Science, University of Chicago, 2004, 5, 11.

29. "Evolution in Florida," *Tampa Morning Tribune*, January 13, 1927.

30. Michael H. Mundt, "The Ku Klux Klan's 1926 Assault on the Democratic Party in Hillsborough County, Florida," *Tampa Bay History* 19, no. 1 (1997): 9; "Stalnaker Surprises Lakeland," *St. Petersburg Evening Independent*, October 22, 1927.

31. "Hillsborough Solon Would Bar Evolution," *St. Petersburg Evening Independent*, December 16, 1926; Associated Press, "Tampan Offers Evolution Bill," *St. Petersburg Times*, April 12, 1927.

32. Florida House of Representatives, *Journal*, 1927, 137.

33. Mary Duncan France, "'A Year of Monkey War': The Anti-evolution Campaign and the Florida Legislature," *Florida Historical Quarterly* 54, no. 2 (1975): 161.

34. "Rollins Faculty Takes Firm Stand against Proposed Evolution Bill," *Rollins Sandspur*, May 6, 1927.35. Associated Press, "Anti-evolution Bill Passed by Committee; Attacked at Hearing," *Tampa Morning Tribune*, April 20, 1927; France, "'Year of Monkey War,'" 165; Associated Press, "Anti-evolution Bill Passed."

36. Associated Press, "Anti-evolution Bill Passed."

37. Associated Press, "Death Knell to Dry Bill," *St. Petersburg Evening Independent*, April 29, 1927.

38. Associated Press, "Principals Amend Proposal against Evolution Measure," *Tampa Morning Tribune*, April 16, 1927; "Doctors Here Send Protest to Capital on Evolution Bill," *Tampa Morning Tribune*, April 26, 1927.

39. "Tampans Organize to Oppose Legislation against Evolution," *Tampa Morning Tribune*, April 29, 1927; "Evolution," advertisement, *Tampa Morning Tribune*, May 4, 1927.

40. Associated Press, "House Dockets Evolution Act," *St. Petersburg Times*, May 7, 1927.

41. Associated Press, "Death Blow to Evolution Bill Dealt by Action of Five House Committees," *Tampa Morning Tribune*, May 12, 1927.

42. Associated Press, "House Evolution Measure Is Killed by Committees," *St. Petersburg Times*, May 12, 1927; France, "'Year of Monkey War,'" 166.

43. Associated Press, "Evolution Phoenix Plunges House into State of Chaos," *St. Petersburg Times*, May 14, 1927; Associated Press, "House Revives Evolution Bill over Committee," *Tampa Morning Tribune*, May 14, 1927.

44. Associated Press, "House Revives Evolution Bill over Committee."

45. Florida House of Representatives, *Journal*, 1927, 3001.

46. Ibid., 3002.

47. Ibid., 3377.

48. Ibid., 5082–83.

49. Associated Press, "Drainage Bill One of Biggest Pieces of Work," *St. Petersburg Times*, June 13, 1927.

50. Associated Press, "Evolution Measure Given Last Page on Senate's Calendar," *Tampa Morning Tribune*, May 26, 1927.

51. Florida House of Representatives, *Journal*, 1927, 4598.

52. Associated Press, "Senate Bars Editor for Rap in Paper on Book Investigation," *Tampa Morning Tribune*, June 3, 1927.

53. Robin Jeanne Sellers, *Femina Perfecta: The Genesis of Florida State University* (Tallahassee: Florida State University Foundation, 1995), 144.

54. Ibid., 145.

55. Ibid., 147.

56. Associated Press, "Objectionable Books Removed," *Tampa Morning Tribune*, November 18, 1927.

57. Sellers, *Femina Perfecta*, 146, 149.

58. Associated Press, "Two Indicted in Fight Made on Textbooks," *Tampa Morning Tribune*, May 19, 1929.

59. "Catts Arrives Here on Tour," *St. Petersburg Times*, December 1, 1927; "To the Thousands of Catts Supporters in Florida," broadside, State Library of Florida Broadsides and Advertisements Collection, ca. 1928, accessed May 6, 2012, http://ibistro.dos.state.fl.us/uhtbin/cgisirsi/?ps=yxPTHoUoi9/STA-FLA/115710161/9.

60. Associated Press, "Text Book Bill Stirs Evolution Talk in Senate," *Tampa Morning Tribune*, May 25, 1929.

61. Florida House of Representatives, *Journal*, 1933, 186.

Chapter 2. "Un-American, Atheistic, Subversive, and Communistic"

1. Florida Department of Education, *A Brief Guide to the Teaching of Science in the Secondary Schools: A Science Teacher Looks at Florida*, Bulletin no. 8 (Tallahassee: Florida Department of Education, 1948), 81.

2. Ibid., 86.

3. Florida Department of Education, *A Guide: Teaching Moral and Spiritual Values in Florida Schools*, Bulletin no. 14 (Tallahassee: Florida Department of Education, 1962), i–ii.

4. Ibid., 58.

5. Ibid., 57.

6. Ibid., 58.

7. Biological Sciences Curriculum Study, *The BSCS Story: A History of the Biological Sciences Curriculum Study* (Colorado Springs: BSCS, 2001), 11.

8. School Board of Dade County, Florida, Minutes of Special Session Meeting, June 21, 1961, 23.

9. Joan Bucks, "Darwin's Monkey on Their Backs," *Miami News*, October 7, 1962; William Tucker, "Repeat of Famed Scopes Drama Scheduled Here," *Miami News*, February 10, 1963.

10. School Board of Dade County, Florida, Minutes of Regular Meeting, November 7, 1962, 7; "He'll Go to Court to Fight Evolution," *Miami News*, December 6, 1962; Tucker, "Repeat of Famed Scopes Drama."

11. School Board of Dade County, Florida, Minutes of Regular Meeting, December 19, 1962, 3, and School Board of Dade County, Florida, Minutes of Regular Meeting, November 7, 1962, 7.

12. School Board of Dade County, Florida, Minutes of Regular Meeting, January 9, 1963, 16.

13. Tucker, "Repeat of Famed Scopes Drama."

14. Ibid.

15. AP report, *Clearwater Sun*, July 20, 1971, quoted in Clarence Elihu Winslow to Pinellas County Board of Education, August 25, 1971, 1.

16. Winslow to Pinellas County Board of Education, August 25, 1971, 1.

17. Clarence Elihu Winslow, "The Bible and Evolution: The Winslow Resolution," in Minutes of Special Meeting, School Board of Pinellas County, Florida, August 25, 1971, 20.

18. School Board of Pinellas County, Florida, Minutes of Special Meeting, August 25, 1971, 19.

19. Florida House of Representatives, *Journal of the House of Representatives*, 1972,

64, http://www.myfloridahouse.gov/Sections/Search/HistoricalJournal/Historical-Journal.aspx.

20. Patti Bridges, "Crusade Asks Schools Tell Bible's Side of Evolution," *St. Petersburg Evening Independent*, January 7, 1972.

21. Ibid.

22. "Televised Debate Draws Overflow Crowd," *Acts and Facts*, April 1974.

23. Karen DeYoung, "Another Skirmish in Great Debate," *St. Petersburg Times*, March 9, 1974.

24. Ibid.

25. Ibid.

26. "1500 Hear Ft. Lauderdale Creation-Evolution Debate," *Acts and Facts*, May 1974.

27. "Evolution, Creation Talk Set," *St. Petersburg Evening Independent*, August 23, 1975.

28. Ann Weldon, "Alternative to Evolution: Schools to Use Creationist Text," *St. Petersburg Evening Independent*, August 30, 1975.

29. Creation Research Society, *Biology: A Search for Order in Complexity* (Grand Rapids, Mich.: Zondervan, 1970), xx.

30. Helen Huntley, "Speakers Plead for Black Studies, 'Creationism,'" *St. Petersburg Times*, August 28, 1975.

31. School Board of Manatee County, Florida, Minutes of Regular Meeting, October 20, 1976, 3.

32. Betty Kohlman, "Minister Again Argues for Religion in School," *St. Petersburg Times*, May 17, 1979.

33. Florida House of Representatives, *Journal*, 1979, 4–5.

34. Christian C. Young and Mark A. Largent, *Evolution and Creationism: A Documentary and Reference Guide* (Westport, Conn.: Greenwood Press, 2007), 229.

35. David Powell, "Bible-Like Evolution Theory May Brew Legislative Clash," *St. Petersburg Times*, November 23, 1979.

36. Ibid.

37. School Board of Hillsborough County, Florida, Minutes of Regular Meeting, December 4, 1979, 468.

38. Ibid.

39. Ibid., 469.

40. Ibid.

41. Ibid.

42. Ibid.

43. Ibid., 470.

44. Ibid.

45. Frank DeLoache, "Ministers, Biology Teacher Ask Schools to Add Creation Theory," *St. Petersburg Times*, December 13, 1979.

46. Ibid.

47. School Board of Hillsborough County, Florida, Minutes of Regular Meeting, December 18, 1979, 503.

48. Ibid., 503, 504.

49. Ibid., 504.

50. Ibid., 505, 506.

51. Ibid., 506, 507.

52. School Board of Hillsborough County, Florida, Minutes of Regular Meeting, March 18, 1980, 80.

53. School Board of Hillsborough County, Florida, Minutes of Regular Meeting, April 22, 1980, 143.

54. Ibid., 144–48.

55. Ibid., 148.

56. Ibid.

57. Ibid.

58. Ibid., 149.

59. School Board of Hillsborough County, Florida, Minutes of Regular Meeting, April 29, 1980, 152.

60. Ibid., 153.

61. Ibid., 154.

62. Ibid., 154, 155.

63. Ibid., 155.

64. Ibid., 156.

65. Ibid.

66. Ibid.

67. Linda Haase, "Creation Theory Vote Draws Great Applause," *Tampa Times*, April 30, 1980.

68. Kenneth M. Pierce, J. Madeleine Nash, and D. L. Coutu, "Education: Putting Darwin Back in the Dock," *Time*, March 16, 1981, 81.

69. Sherry Howard, "Board OKs Theory of Creationism," *Tampa Tribune*, April 30, 1980.

70. Jon Peck and Linda Haase, "Creationism Bid in House May Get Boost," *Tampa Times*, May 5, 1980; Florida House of Representatives, *Journal*, 1980, 17.

71. Peck and Haase, "Creationism Bid in House"; Howard, "Board OKs Theory of Creationism."

72. Peck and Haase, "Creationism Bid in House."

73. Florida House of Representatives, *Journal*, 1980, 473.

74. School Board of Manatee County, Florida, Minutes of Regular Meeting, June 17, 1980, 12.

75. School Board of Manatee County, Florida, Minutes of Regular Meeting, September 16, 1980, 2.

76. Ibid.

77. Betty Kohlman, "School Board to Okay Curriculum before Scientific Creation Theory Is Taught," *St. Petersburg Times*, October 1, 1980.

78. Marilyn Brown, "Permission to Teach Creation Sought," *St. Petersburg Evening Independent*, September 24, 1980.

79. School Board of Pinellas County, Florida, Minutes of Regular Meeting, September 24, 1980, 2; Charla Wasel, "Creationism Issue Divides Pinellas," *St. Petersburg Evening Independent*, November 15, 1980.

80. Wasel, "Creationism Issue Divides Pinellas."

Chapter 3. "A Spirit of Compromise and Conciliation"

1. John Heagney, "Creationism in Pasco Schools?" *St. Petersburg Times*, February 3, 1981.

2. John Heagney, "Board Delays Creationism Decision," *St. Petersburg Times*, February 4, 1981.

3. Heagney, "Creationism in Pasco Schools?"; Heagney, "Board Delays Creationism Decision."

4. Heagney, "Board Delays Creationism Decision."

5. Laurin A. March, "School Board Again Debates Teaching 'Scientific Creationism,'" *St. Petersburg Times*, June 4, 1981.

6. John Heagney, "Book on Creationism May Be Placed in Pasco County School Libraries," *St. Petersburg Times*, November 4, 1981.

7. School Board of Manatee County, Florida, Minutes of Regular Meeting, March 17, 1981, 4.

8. School Board of Manatee County, Florida, "Proposed Policy on Creation vs. Evolution," Supplemental Minute File, November 17, 1981.

9. School Board of Manatee County, Florida, Minutes of Regular Meeting, November 17, 1981, 7, 6, 7.

10. Gus Sakkis to School Board of Pinellas County, memorandum, February 11, 1981, Nancy Marsh personal archive.

11. "Okaloosa Schools May Teach Creation Theory," *Ocala Star-Banner*, March 11, 1981; "Okaloosa School Superintendent Backs Off on Creationism," *Lakeland Ledger*, July 29, 1981.

12. Robin Williams, "Churchwell Not Opposed to Teaching Creationism," *Lakeland Ledger*, March 14, 1981.

13. Darlene Williams, "Citizens for Morality Schedule 1st Meeting; Pastor to Give Speech," *Ocala Star-Banner*, March 29, 1981.

14. Cherie Beers, "Differences Expressed on Creationism Ruling," *Ocala Star-Banner*, January 10, 1982.

15. Ibid.

16. K. W. Goodman, "'Instant Creation' Theory May Be Included in Courses," *Daytona Beach Morning Journal*, March 14, 1980.

17. Ibid.

18. "Science Instructors Oppose Creationism," *Daytona Beach Morning Journal*, May 21, 1980.

19. "'Creationism' Bill Loses in the Senate," *Palm Beach Post*, June 5, 1981.

20. Nancy Marsh, e-mail to author, April 21, 2009.

21. Nancy Marsh to Sam Horton, "Origins Curriculum Development," inter-office communication, December 30, 1981, Nancy Marsh personal archive.

22. Wayne Moyer to Nancy Marsh, May 7, 1980, Nancy Marsh personal archive.

23. "Should Creation Have Equal Time with Evolution in Science Classes?" *Awake*, September 22, 1981, 27.

24. Betty Hoskins to Nancy Marsh, November 8, 1980, Nancy Marsh personal archive.

25. John Betz, "Look What They're Teaching Our Children," *Tampa Magazine*, September 1981, 30.

26. "Final Origins Curriculum Guidelines," n.d., Nancy Marsh personal archive; and Gil Klein, "Putting 'Creation' in the Curriculum: Tampa Agrees to Do It—but How?" *Christian Science Monitor*, March 30, 1981.

27. "Final Origins Curriculum Guidelines," n.d., Nancy Marsh personal archive.

28. Martin Sandberg, letter, n.d., Nancy Marsh personal archive.

29. John Betz, Thomas Bowers, Randy Akers, Henry Mushinsky, Richard Taylor, and H. Edwin Steiner to the Hillsborough County School Board, "Partial Dissent from the Origins Curriculum Committee guidelines and Goals for Discussion and Instruction," December 2, 1980, Nancy Marsh personal archive.

30. Ibid.

31. Ibid.

32. Sherry Howard, "Board OKs Creationism Guidelines," *Tampa Tribune*, December 17, 1980.

33. Ibid.

34. Sam Horton to W. Crosby Few, October 1, 1980, Nancy Marsh personal archive.

35. W. Crosby Few to Sam Horton, November 5, 1980, Nancy Marsh personal archive.

36. G. R. Jimenez to Sam Horton, "Report Dealing with Social Studies Origin Committee Recommended Guidelines," inter-office communication, April 29, 1981, Nancy Marsh personal archive.

37. William Mayer to Nancy Marsh, April 8, 1981, Nancy Marsh personal archive.

38. Nancy Marsh, "Scientific Creationist to Speak 3/16/81 in Bradenton," inter-office communication, March 4, 1981, Nancy Marsh personal archive; and Nancy Marsh, "Approval for an Out-of-County Teacher Trip for Staff Development Purposes," inter-office communication, March 6, 1981, Nancy Marsh personal archive.

39. Patti Breckenridge, "A Matter of Degree: Creationists Create New Title for Darwin," *Tampa Tribune*, January 5, 1982.

40. Hillsborough County Schools, "Preliminary Draft: The Study of Origins," n.d., Nancy Marsh personal archive, 3, 6.

41. Ibid., 6.

42. Ibid., 37.

43. Ibid., 43, 44.

44. School Board of Hillsborough County, Florida, Minutes of Meeting, January 11, 1982, 20.

45. Ibid.

46. Ibid.

47. Ibid., 21.

48. Ibid., 22.

49. Ibid.

50. Bruce McKay, *Science Research Proves Evolution Hoax: The Conflagration, When Parallel Universes Merge* (Bloomington, Ind.: AuthorHouse, 2006).

51. School Board of Hillsborough County, Florida, Minutes of Meeting, January 11, 1982, 22.

52. Ibid., 23.

53. Ibid., 23–24.

54. Ibid., 24.

55. Ibid.

56. Ibid., 25.

57. Ibid.

58. Ibid., 25–26.

59. School Board of Hillsborough County, Florida, Minutes of Meeting, January 19, 1982, 43–44.

Chapter 4. "A History of Hoaxes, Deception, and Deceit"

1. School Board of Manatee County, Florida, Minutes of Regular Meeting, June 7, 1983, 3.

2. Ibid.

3. Julie Ross, "Creationist Teaching Urged for Schools," *Sarasota Herald-Tribune*, June 30, 1983.

4. School Board of Manatee County, Florida, Minutes of Regular Meeting, June 29, 1983, 5.

5. Ibid., 6.

6. Julie Ross, "Johnson Wants Equal Time for Creationism," *Sarasota Herald-Tribune*, July 12, 1983.

7. School Board of Manatee County, Florida, Minutes of Regular Meeting, August 30, 1983, 1.

8. School Board of Manatee County, Florida, Minutes of Regular Meeting, September 6, 1983, 1.

9. Julie Ross, "School Board Holds Stance on Religion," *Sarasota Herald-Tribune*, September 7, 1983.

10. School Board of Manatee County, Florida, Minutes of Regular Meeting, September 6, 1983, 2, 3.

11. Ibid., 3.

12. Ross, "School Board Holds Stance."

13. Julie Ross, "Court Rulings Put Creationism Fight into Perspective," *Sarasota Herald-Tribune*, September 11, 1983.

14. Ibid.

15. School Board of Manatee County, Florida, Minutes of Regular Meeting, September 6, 1983, 6.

16. Ross, "School Board Holds Stance."

17. Ibid.

18. "The Student Science Research Program," advertisement, *Sarasota Herald-Tribune*, September 12, 1983.

19. Fred M. Hechinger, "About Education: Textbooks Criticized as Simplistic and Dull," *New York Times*, April 10, 1984.

20. Ibid.

21. Ellie McGrath, "Education: Texas Eases Up on Evolution," *Time*, April 30, 1984.

22. Associated Press, "Book-Buying Alliance Falters," *Palm Beach Post*, March 21, 1984.

23. M. C. Poertner, "Tussling over Textbooks: Mediocre Books Result from Political Grappling," *Orlando Sentinel*, October 27, 1985.

24. Mark Potok, "Broward Adds Voice to Debate over Evolution," *Miami Herald*, October 20, 1985.

25. Herbert A. Smith, Ralph P. Frazier, and Michael A. Magnoli, *Exploring Living Things*, 2nd ed. (River Forest, Ill.: Laidlaw Brothers, 1980), 4.

26. Ibid., 491, 527.

27. Associated Press, "Turlington Defends Book Selection," *Lakeland Ledger*, October 14, 1975.

28. Bill Kaczor, "Cabinet Wants New Bidding for Teacher-Testing Contract," *St. Petersburg Times*, March 19, 1980; Ike Flores, "Watchdog Hounds State Schools," *St. Petersburg Evening Independent*, August 14, 1981; Bill Kaczor, "Cabinet Approves Single Calendar for Universities," *Sarasota Herald-Tribune*, February 23, 1980; Bill Kaczor, "Cabinet Wants New Bidding for Teacher-Testing Contract," *St. Petersburg Times*, March 19, 1980.

29. Ellyn Ferguson, "Family Lobbyist Continues Fight against School System's Humanism," *Lakeland Ledger*, April 3, 1980.

30. Marilyn A. Moore, "Woman Fights Textbooks That 'Reprogram' Youth," *Palm Beach Post*, May 8, 1983.

31. Tyler Ward, "Humanists in Schools Linked to Corruption," *Ocala Star-Banner*, May 6, 1981. Ellipsis in original.

32. Tyler Ward, "Humanists in Schools Linked to Corruption," *Ocala Star-Banner*, May 6, 1981. Ellipsis in original.

33. M. C. Poertner, "Tussling over Textbooks."

34. Associated Press, "Evolution Texts Draw Criticism from Educators," *Miami Herald*, February 19, 1986.

35. Ibid.

36. Joe Bizzaro, "Cabinet Delays OK of 23 Disputed Texts," *Palm Beach Post*, February 19, 1986.

37. Ibid.; Associated Press, "Evolution Texts Draw Criticism."

38. United Press International, "Creationists Win Book Delay," *Sun-Sentinel* (South Florida), February 19, 1986; Associated Press, "Evolution Texts Draw Criticism"; Robert Barnes, "Graham: Creationism, Evolution Both Could Be Taught in Schools," *St. Petersburg Times*, February 21, 1986.

39. Linda Kleindienst, "Evolution Theory Texts OK'd Despite Protests," *Sun-Sentinel*, March 5, 1986.

40. Ibid.; Paul Anderson, "23 Controversial Texts Approved over Protests," *Miami Herald*, March 5, 1986.

41. Ed Birk, "Creationists Lose Book Debate," *Lakeland Ledger*, March 5, 1986; Charles Holmes, "Creationists Lose Fight over Evolution Teaching," *Palm Beach Post*, March 5, 1986.

42. Donna Blanton, "Creationists Angry over Textbook List," *Orlando Sentinel*, March 5, 1986.

43. Ibid.; Don Horine, "Fundamentalists' Censorship Leading Own Road of Success," *Palm Beach Post*, August 4, 1986.

44. Birk, "Creationists Lose Book Debate."

45. School Board of Escambia County, Florida, Minutes of Regular Meeting, May 27, 1986, 11–12.

46. Associated Press, "Group Wants Creation Taught," *St. Petersburg Times*, July 21, 1986.

47. Ibid.

48. Associated Press, "School Board Rejects Creationism Handouts," *Orlando Sentinel*, July 24, 1986; Associated Press, "Group Wants Creation Taught," *St. Petersburg Times*, July 21, 1986.

49. Associated Press, "Presentation of Evolution Disputed in Pensacola Schools," *St. Petersburg Evening Independent*, July 21, 1986.

50. Ibid.; "Board Says No to Creationism Handouts," *St. Petersburg Times*, July 24, 1986.

51. Charles Holmes, "Republican Gubernatorial Hopefuls Support Teaching of Creationism," *Palm Beach Post*, August 16, 1986.

52. Robert Barnes, "GOP Governor Candidates Want Schools to Teach Creationism," *St. Petersburg Times*, August 16, 1986; Linda Kleindienst, "Call to Conservatives," *Sun-Sentinel*, August 18, 1986.

53. Barnes, "GOP Governor Candidates."

54. John Kennedy, "'Morality' Key Issue in Local Political Races," *Sun-Sentinel*, August 24, 1986; Jeffrey Weiss, "Religion Issues Stir Up School Board Races," *Miami Herald*, August 29, 1986.

55. Ken Swart and Lori Crouch, "Religion Shadows School Race," *Sun-Sentinel*, August 10, 1986.

56. Steve Nichol, "Incumbent Battles Grass-Roots Candidate," *Sun-Sentinel*, August 28, 1986.

57. Angie Cannon, "Christian Candidates Ignite School Race," *Miami Herald*, October 5, 1986.

58. Tom Scherberger, "Pajcic, Martinez Trade Digs over Budget, Environment," *Orlando Sentinel*, October 17, 1986.

59. Ed Birk, "Textbooks Approved Despite Creationists," *Lakeland Ledger*, February 18, 1987.

60. Ibid.

61. Associated Press, "Group May File Suit over Sagan Book's Use," *Gainesville Sun*, November 19, 1987.

62. United Press International, "Anti-evolution Policy Rejected," *Orlando Sentinel*, January 16, 1988.

63. Ibid.

Chapter 5. "A Conspiracy to Destroy the Faith of Children"

1. Karen Brandon, "Teacher Reprimanded for Film on Creationism," *Sun-Sentinel*, March 22, 1988; Kimberly Crockett, "Creationism Lesson Rebuked," *Miami Herald*, March 22, 1988; Brandon, "Teacher Reprimanded."

2. Crockett, "Creationism Lesson Rebuked."

3. Brandon, "Teacher Reprimanded."

4. James D. Davis, "Creationists Say They Have Just Begun to Fight," *Sun-Sentinel*, June 18, 1988; Karen Brandon, "Teachers Can Discuss Creationism in Class," *Sun-Sentinel*, April 7, 1988.

5. Davis, "Creationists Say They Have Just Begun."

6. Ibid.

7. School Board of Manatee County, Florida, Minutes of Regular Meeting, February 21, 1989, 1; School Board of Manatee County, Florida, Minutes of Regular Meeting, April 18, 1989, 1.

8. School Board of Manatee County, Florida, Minutes of Regular Meeting, June 6, 1989, 4; School Board of Manatee County, Florida, Minutes of Regular Meeting, October 17, 1989, 1.

9. School Board of Manatee County, Florida, Minutes of Regular Meeting, January 2, 1990, 1.

10. School Board of Pinellas County, Florida, Minutes of Regular Meeting, March 14, 1990, 1; School Board of Manatee County, Florida, Minutes of Regular Meeting, September 6, 1990, 1.

11. Edwards v. Aguillard, 482 U.S. 578 (1987), *FindLaw*, accessed June 29, 2012, http://laws.findlaw.com/us/482/578.html.

12. Percival Davis and Dean H. Kenyon, *Of Pandas and People: The Central Question of Biological Origins*, 2nd ed. (Dallas: Haughton, 1996), 26.

13. Center for Science and Culture, "What Is Intelligent Design?" Discovery Institute, accessed March 24, 2013, http://www.intelligentdesign.org/whatisid.php.

14. Erik Larson, "Darwinian Struggle: Instead of Evolution, a Textbook Proposes 'Intelligent Design,'" *Wall Street Journal*, November 14, 1994.

15. Ellen Moses, "Reverend Stresses Evolution by Design," *Bradenton Herald*, October 21, 1990; School Board of Manatee County, Florida, Minutes of Regular Meeting, September 6, 1990, 1.

16. School Board of Manatee County, Florida, Minutes of Regular Meeting, October 2, 1990, 1.

17. School Board of Manatee County, Florida, Minutes of Regular Meeting, November 7, 1990, 3.

18. Ibid.

19. School Board of Manatee County, Florida, Minutes of Regular Meeting, September 3, 1991, 1.

20. School Board of Manatee County, Florida, Minutes of Regular Meeting, February 4, 1992, 3, 4.

21. "Obituaries," *St. Petersburg Times*, July 20, 2002.

22. Morris Kennedy, "Martinez Supports 10th State University, Teaching Creationism," *Tampa Tribune*, September 21, 1990.

23. Morris Kennedy, "Creationism Flawed Belief, MacKay Says," *Tampa Tribune*, September 25, 1990.

24. Ibid.

25. "Candidates Asked to Retract Statements," *Tampa Tribune*, September 29, 1990; "Monkeying Around," *Palm Beach Post*, September 30, 1990.

26. Linda Kleindienst, "Martinez Ad Hits MacKay on Creation," *Sun-Sentinel*, October 2, 1990.

27. Charles Holmes, "Creationism Ignites Race for Governor," *Palm Beach Post*, October 2, 1990.

28. Ibid.; Morris Kennedy, "Martinez Unloads with 3 New Ads," *Tampa Tribune*, October 2, 1990.

29. "Most Callers Want Creationism Taught in Public Schools," *Orlando Sentinel*, October 2, 1990; Associated Press, "Poll: Most for Teaching Creationism," *Sun-Sentinel*, October 20, 1990.

30. John C. Van Gieson, "MacKay Apologizes for Remarks on Creationism," *Orlando Sentinel*, October 6, 1990.

31. John D. McKinnon, "MacKay Apologizes for Remark about Creationism," *St. Petersburg Times*, October 6, 1990.

Chapter 6. "It Was Historic, Wasn't It?"

1. Rick Badie, "Cowin, Hart Differ in Their Approach to School System," *Orlando Sentinel*, August 31, 1990; Rick Badie, "Cowin Won't Quit Fighting for Schools," *Orlando Sentinel*, November 13, 1990.

2. "Re-elect Members of Lake County School Board," *Orlando Sentinel*, August 12, 1990; Jeff Brazil, "200 Participate in Anti-abortion Rally in Tavares," *Orlando Sentinel*, October 10, 1989; and Bill Boyd, "2 from Lake in Utah for Assembly," *Orlando Sentinel*, June 2, 1990; "New to CWA?" Concerned Women for America, accessed July 11, 2012, http://www.cwfa.org/about.asp.

3. Jim Runnels, "Cowin Is Defeated by 12 Votes," *Orlando Sentinel*, September 6, 1990; Rick Badie, "School Activists Seeking More Attention to Religion," *Orlando Sentinel*, April 7, 1991; Rick Badie, "Hart Says Textbook Backs Promiscuity," *Orlando Sentinel*, November 29, 1990.

4. Badie, "Hart Says Textbook Backs Promiscuity"; Badie, "School Activists Seeking More Attention."

5. "Part of Board Likes Full-Service Schools," *Orlando Sentinel*, October 24, 1991.

6. School Board of Lake County, Florida, Minutes of Regular Meeting, March 26, 1991, 5; Badie, "School Activists Seeking More Attention"; Bob Wells, handwritten request to Lake County School Board, April 8, 1991.

7. Linda Rozar to Lake County School Board, April 22, 1991.

8. Badie, "School Activists Seeking More Attention."

9. Rick Badie, "Hart Wins Backing on Move for Hearing," *Orlando Sentinel*, April 25, 1991.

10. School Board of Lake County, audio recording of regular school board meeting, April 23, 1991, cassette tapes, tape 1, side A.

11. Ibid.

12. Ibid.

13. Ibid.

14. Ibid.

15. Ibid.

16. Ibid., tape 1, side B.

17. Ibid., tape 1, side A.

18. Ibid.

19. Ibid., tape 1, side B.

20. Ibid.

21. Ibid.

22. Ibid.

23. Ibid.

24. Ibid.

25. Ibid.

26. Ibid., tape 2, side A.

27. Ibid.

28. Rick Badie, "School Board to Discuss Theory," *Orlando Sentinel*, April 25, 1991; Badie, "Hart Wins Backing of Move."

29. Badie, "School Board to Discuss Theory."

30. Bill Trask, "Lake School Battle," *Daily Commercial*, May 12, 1991; Badie, "School Board to Discuss Theory"; "Most Callers Want Creationism Taught in Public Schools," *Orlando Sentinel*, April 30, 1991.

31. Rick Badie, "Voices Join Cry for Creationism Debate," *Orlando Sentinel*, May 5, 1991.

32. Ibid.; Rick Badie, "Creationism in Schools: Union Votes No," *Orlando Sentinel*, May 7, 1991.

33. Rick Badie, "Foes Line Up to Fight Creationism Groups," *Orlando Sentinel*, May 10, 1991.

34. Rick Badie, "Lobbying Efforts in Full Swing," *Orlando Sentinel*, May 12, 1991; Trask, "Lake School Battle."

35. Badie, "Lobbying Efforts in Full Swing."

36. Ibid.

37. "School Board Members' Views," *Orlando Sentinel*, May 12, 1991.

38. School Board of Lake County, audio recording of school board workshop, May 13, 1991, cassette tapes, tape 1, side A.

39. Ibid.

40. Ibid.

41. Ibid.

42. Ibid.

43. Ibid.

44. Ibid.

45. Ibid., tape 1, side B.

46. Ibid.

47. Ibid.

48. Ibid., tape 2, side A.

49. Ibid.

50. Ibid., tape 2, side B.

51. Ibid.

52. Ibid.

53. Ibid.

54. Rick Badie, "Creationism Fight Continues to Evolve," *Orlando Sentinel*, May 16, 1991.

55. School Board of Lake County, audio recording of regular school board meeting, May 14, 1991, cassette tape, tape 1, side A.

56. Ibid.

57. Ibid.

58. Ibid.

59. Ibid.

60. Ibid.

61. Ibid.

62. Ibid.

63. Ibid.

64. Linda Chong, "Minister Forms Plan to Teach Creationism," *Orlando Sentinel*, June 1, 1991.

Chapter 7. "One of the Primal Evils in Our Country"

1. Rick Tonyan, "Creationists Still Working on Getting Their Theory Taught in Science Class," *Orlando Sentinel*, January 13, 1991.

2. Ibid.

3. Beth Taylor, "Independence Costs Stetson," *Orlando Sentinel*, February 11, 1993.

4. Beth Muniz, "Ministerial Group: Creationist Book Needed in Schools," *Bradenton Herald*, January 6, 1991.

5. Ibid.

6. Carlos Galarza, "Creation Debate Heats Up Locally," *Bradenton Herald*, December 9, 1991.

7. Ibid.

8. Ibid.

9. Madrigal, e-mail to author, September 6, 2011; Carlos Galarza, "Evolution/Creationism Debate Stirs Passions at Civic Center," *Bradenton Herald*, January 12, 1992.

10. Galarza, "Evolution/Creationism Debate Stirs Passions"; Madrigal, e-mail to author.

11. Madrigal, e-mail to author.

12. Ibid.

13. Kim Dutra, "Professors Rap Local Teachers' Credits for Creationism Debate," *Bradenton Herald*, February 22, 1992.

14. Kathleen Beeman, "Professors Challenge Creationism Seminar," *Tampa Tribune*, February 25, 1992; Dutra, "Professors Rap Local Teachers' Credits."

15. Beeman, "Professors Challenge Creationism Seminar."

16. Susan Jacobson, "Parent Wants Schools to Teach Creationism," *Orlando Sentinel*, May 16, 1993.

17. Ibid.

18. Susan Jacobson, "Creation in Schools Fails to Fly," *Orlando Sentinel*, May 28, 1993.

19. Ibid.

20. Florida House of Representatives, *Journal of the House of Representatives*, 1994, 496, http://www.myfloridahouse.gov/Sections/Search/HistoricalJournal/Historical-Journal.aspx; "Bashing Fruit? Watch Language," *Sun-Sentinel*, March 25, 1994.

21. Anne Lindberg, "Most School Candidates Oppose Teaching Creationism," *St. Petersburg Times*, August 31, 1994.

22. Lesley Collins, "Hopefuls' Church-State Views Vary," *Tampa Tribune*, September 5, 1994; Anne Lindberg, "2nd Hopeful Backs Teaching of Creationism," *St. Petersburg Times*, September 2, 1994.

23. Dave Weber, "McBrady, Wiseman Run in Different Directions," *Orlando Sentinel*, October 1, 1994.

24. Larry Barszewski, "Labels Hang on Hopefuls," *Sun-Sentinel*, October 16, 1994; Ledyard King, "School Foes Trade Barbs at Forum," *Palm Beach Post*, November 1, 1994.

25. Barszewski, "Labels Hang on Hopefuls."

26. Ibid.

27. Mary Lou Pickel, "St. Lucie School Board Unaware of Lobbying for Creationism," *Palm Beach Post*, January 20, 1995; Jacques Picard, "Meaning of Creationism Is Open to Interpretation," *Fort Pierce Tribune*, January 19, 1995.

28. "Creating a Diversion," *Palm Beach Post*, January 24, 1995.

29. Jacques Picard, "In the Beginning: Theories of Creationism, Evolution Clash in Classroom," *Fort Pierce Tribune*, January 19, 1995.

30. Ibid.

31. Jacques Picard, "Creationism Textbook Already May Be in Use," *Fort Pierce Tribune*, January 20, 1995.

32. Picard, "In the Beginning."

33. Ibid.; Mary Lou Pickel, "St. Lucie Schools Panel Reviewing Creation Text," *Palm Beach Post*, January 19, 1995.

34. Jacques Picard, "Evolution-Creationism Debate Creates Friction," *Fort Pierce Tribune*, January 29, 1995; Picard, "Creationism Textbook."

35. Picard, "In the Beginning"; Pickel, "St. Lucie School Board Unaware of Lobbying"; David Mosrie, "Creationism in the Classroom," *Fort Pierce Tribune*, January 29, 1995; Mary Lou Pickel, "Politics vs. Religion Puts Mosrie in Hot Seat Again," *Palm Beach Post*, February 5, 1995.

36. Picard, "Creationism Textbook."

37. School Board of St. Lucie County, Florida, Minutes of Regular Meeting, January 24, 1995, 4.

38. Ibid., 6.

39. Jacques Picard, "Creationism Crops Up at Meeting," *Fort Pierce Tribune*, January 25, 1995.

40. School Board of St. Lucie County, Florida, Minutes of Regular Meeting, January 24, 1995, 6; Picard, "Creationism Crops Up at Meeting."

41. Picard, "Evolution-Creationism Debate."

42. Mosrie, "Creationism in the Classroom."

43. Jacques Picard, "Local Lawmaker Wins Award from Coalition," *Fort Pierce Tribune*, February 2, 1995; Tonya Tulloss, "Religious Coalition's Director Quits Post," *Fort Pierce Tribune*, November 11, 1995.

44. "'A' for Brogan on Talk, but Just a 'C' for Ideas," *Palm Beach Post*, January 15, 1995.

45. Ron Matus, "History Lesson on Science Standards," *The Gradebook* (blog), *St. Petersburg Times*, November 28, 2007, http://www.tampabay.com/blogs/gradebook/2007/11/history-lesson.html.

46. Barbara Forrest, "Unmasking the False Prophet of Creationism," *Reports of the National Center for Science Education* 19, no. 5 (September–October 1999), http://ncse.com/rncse/19/5/unmasking-false-prophet-creationism; Andrew Petto, Stephen Meyers, and Bob Leipold, "Dr Dino Does 'Delphia,'" *Reports of the National Center for Science Education* 19, no. 5 (September–October 1999), http://ncse.com/rncse/19/5/dr-dino-does-delphia.

47. Greg Martinez, "Stupid Dino Tricks: A Visit to Kent Hovinds's Dinosaur Adventure Land," *Skeptical Enquirer* 28, no. 6 (November–December 2004): 47–51; Abby Goodnough, "Darwin-Free Fun for Creationists," *New York Times*, May 1, 2004.

48. Herald Wire Services, "IRS Investigating Owner of Creationist Museum," *Miami Herald*, April 18, 2004; Nicole Lozare, "'Dr. Dino' Guilty on All Counts," *Pensacola News Journal*, November 3, 2006.

49. Kris Wernowsky, "Judge Clears Way for Dinosaur Park to Be Seized," *Pensacola News Journal*, August 1, 2009; "Group Tours," The Creation Store, accessed July 26, 2012, http://pensacola.creationstore.org/birthdays.php.

50. Bob Massey, "Speaker to Attack Theory of Evolution," *North Port Sun*, September 9, 2000.

51. "About Creation Studies Institute," Creation Studies Institute, accessed July 26, 2012, http://www.creationstudies.org/aboutus.html; "Ice Age Fossil Adventure," Creation Studies Institute, accessed July 26, 2012, http://www.creationstudies.org/htdocs/fossilfloat.html.

52. Lona O'Connor, "Parents Exercise Theory of Choice," *Sun-Sentinel*, May 24, 1998.

53. Gary Parker and Mary Parker, "Creation Education Vacation Cover Letter 2011," accessed July 26, 2012, http://www.creationadventuresmuseum.org/docs/CEV_Cover_Letter_2011.pdf; Lona O'Connor, "Bones Tell the Story," *Sun-Sentinel*, January 27, 2001.

54. Michelle Bearden, "Scientist Evolves into Creationist," *Tampa Tribune*, September 23, 2000.

55. Bob Phelps, "Touting Creation Evidence," *Florida Times-Union*, March 27, 1998.

56. Elliott Jones, "Answers in Genesis," *Vero Beach Press Journal*, December 14, 1998.

Chapter 8. "There Are Razor Blades in That Apple"

1. Eileen Kelley, "Lee School Board Considers Adding Bible as Literature Class," *Naples Daily News*, January 19, 1996.

2. Ibid.

3. Eileen Kelley, "Committee OKs Bible for School," *Naples Daily News*, February 23, 1996.

4. Kelley, "Lee School Board Considers Adding Bible"; Kelley, "Committee OKs Bible for School."

5. Eileen Kelley, "Lee School Board OKs Bible Class," *Naples Daily News*, March 27, 1996; Eileen Kelley, "Bible Class Delayed until 1997–98 School Year," *Naples Daily News*, October 9, 1996.

6. Eileen Kelley, "Lee Warned about Bible Curriculum," *Naples Daily News*, December 18, 1996; Tom Fiedler, "Battle over the Bible Divides Lee County," *Miami Herald*, June 2, 1997.

7. Fiedler, "Battle over the Bible Divided Lee County"; Eileen Kelley, "Beleaguered Attorney Steps Down," *Naples Daily News*, January 8, 1997.

8. Eileen Kelley, "Attorney to Defend Bible Class," *Naples Daily News*, January 22, 1997.

9. Fiedler, "Battle over the Bible"; Robert Steinback, "Will Science Be Sacrificed in Lee School?" *Miami Herald*, June 6, 1997.

10. Eileen Kelley, "Bible Class Makes Grade at Estero High," *Naples Daily News*, May 28, 1997.

11. Eileen Kelley, "Adam, Eve Ousted from Lee Schools," *Naples Daily News*, July 15, 1997; Tom Fiedler, "Courts May Decide How Bible Course Should Be Taught," *Miami Herald*, August 4, 1997.

12. Eileen Kelley, "Lee Approves Bible Curriculum," *Naples Daily News*, August 7, 1997.

13. Eileen Kelley, "Bible out of Lee Schools This Term," *Naples Daily News*, August 12, 1997; Kara Vick, "Lee Bible Dispute Has Been 'Unchristian,'" *Naples Daily News*, January 18, 1998.

14. "Lee County Sued over Bible Classes," *Stuart News*, December 10, 1997.

15. Lisa Holewa, "Court: Bible Class Can Be Taught in Public Schools," *Bradenton Herald*, January 21, 1998; Kara Vick, "One Bible Class OK'd, Other Halted," *Naples Daily News*, January 21, 1998; Kara Vick, "Students Discuss Lawsuits on 1st Day of Bible Classes," *Naples Daily News*, January 22, 1998.

16. Kara Vick, "Official Won't Support Creationism in Lee," *Naples Daily News*, February 7, 1998.

17. Thomas Pear, "School Board Candidates Debate the Issues," *Cape Coral Daily Breeze*, August 26, 1996, quoted in Connie Holzinger, "When Worldviews Collide: The Issue of a Bible History Class in the Lee County Public Schools," *Ampersand*, January 13, 1998, http://itech.fgcu.edu/&/issues/vol1/issue1/worldviews.htm.

18. Kara Vick, "Committee: Creationism Should Be Discussed Too," *Naples Daily News*, January 30, 1998.

19. Kara Vick, "Bible Suit Settled in Lee," *Naples Daily News*, February 26, 1998.

20. Douglas Kalajian, "Lee County Bible Battle Wanes as Christian Group Loses Its Edge," *Palm Beach Post*, September 21, 1998.

21. School Board of Manatee County, Florida, Minutes of Regular Meeting, November 1, 1999, 8.

22. Rod Harmon, "Manatee Pastor Pushes Creationism," *Bradenton Herald*, November 20, 1999.

23. Ibid.

24. Duane Marsteller, "Evolution vs. Creation," *Bradenton Herald*, November 23, 1999.

25. Jennifer Merritt and Jeffrey Solochek, "Creation Argument Launched," *Sarasota Herald-Tribune*, November 28, 1999.

26. Annette Ayres, "Ethics Question Arises from Creationism," *Bradenton Herald*, December 2, 1999.

27. Ibid.; Marsteller, "Evolution vs. Creation."

28. Marsteller, "Evolution vs. Creation"; Merritt and Solochek, "Creation Argument Launched."

29. Ayres, "Ethics Question Arises"; Merritt and Solochek, "Creation Argument Launched."

30. Pat Kelly, "Forums Focus on Area Issues," *Bradenton Herald*, August 23, 2000.

31. Thomas Lee Tryon, "A Value Judgment for Voters," *Sarasota Herald-Tribune*, November 1, 1998.

32. Denise Zoldan, "Revision 6: Making Strong Commitment to Public Schools," *Naples Daily News*, November 1, 1998.

33. Martin Dyckman, "A 'Mandate' for Vouchers? No Way," *St. Petersburg Times*, March 11, 1999; Mark Silva, "Bush Plan for Schools Draws Praise," *Miami Herald*, January 26, 1999.

34. David Pedreira, "Pro-voucher Group Not Short on Cash," *Tampa Tribune*, March 12, 1999.

35. "McKay Scholarship Program," Florida Department of Education, accessed October 4, 2012, http://www.floridaschoolchoice.org/information/mckay/.

36. Florida Department of Education, *FTC Scholarship Program* (Tallahassee: Florida Department of Education, 2012), 1, http://www.floridaschoolchoice.org/Information/CTC/files/FTC_Sept_2012.pdf.

37. Florida House of Representatives, *House Bill 751: Education*, 1999, 23, http://archive.flsenate.gov/data/session/1999/House/bills/billtext/pdf/h0751.pdf; Florida Department of Education, *McKay Scholarship Program* (Tallahassee: Florida Department of Education, 2012), 2, http://www.floridaschoolchoice.org/Information/McKay/files/Fast_Facts_McKay.pdf; Florida Department of Education, *FTC Scholarship Program*, 2.

38. S. V. Date, "Many Private Schools Taking Vouchers Teach Creationism," *Palm Beach Post*, October 9, 2005.

39. Ibid.; A Beka Book, *Observing God's World* (Pensacola: A Beka Book, n.d.), accessed October 4, 2012, https://www.abeka.com/ABekaOnline/BookDescription.aspx?sbn=67024.

40. Florida Senate, *Constitution of the State of Florida*, Article I, Section 3, accessed October 4, 2012, http://www.flsenate.gov/Laws/Constitution#A1S03.

41. "Holmes v. Bush," Americans United, October 24, 2011, http://www.au.org/our-work/legal/lawsuits/holmes-v-bush.

42. "Fordham Mission," Thomas B. Fordham Foundation, n.d., accessed October 4, 2012, http://www.edexcellence.net/about-us/fordham-mission.html; Lawrence S. Lerner, "State Science Standards: An Appraisal of Science Standards in 36 States," *Fordham Report* 2, no. 4 (March 1998): 18, 19.

43. Lawrence S. Lerner, *Good Science, Bad Science: Teaching Evolution in the States* (Washington, D.C.: Thomas B. Fordham Foundation, 2000), 16.

44. Paul R. Gross, *The State of State Science Standards 2005* (Washington, D.C.: Thomas B. Fordham Foundation, 2005), 34.

45. Jodi Wilgoren, "Politicized Scholars Put Evolution on the Defensive," *New York Times*, August 21, 2005; Marc Caputo, "Gov. Bush Oddly Evasive on Evolution," *Miami Herald*, October 9, 2005.

46. Caputo, "Gov. Bush Oddly Evasive on Evolution."

47. Kimberly Miller, "Chancellor's Creationism Views Draw Out Her Critics," *Ocala Star-Banner*, October 9, 2005.

48. Ibid.; Scott Maxwell, "K–12 Chief Holds Strong Views," *Orlando Sentinel*, September 4, 2005.

49. Miller, "Chancellor's Creationism Views"; Ron Matus, "Candidate: Story on Me Is Wrong," *St. Petersburg Times*, June 26, 2007.

50. Matus, "Candidate: Story on Me Is Wrong"; Ron Matus, "Will Darwin Come Under Fire in Florida?" *St. Petersburg Times*, August 28, 2005.

51. Matus, "Will Darwin Come Under Fire in Florida?"

52. Howard Goodman, "Quest for 'Academic Freedom' Could Be Anything But," *Sun-Sentinel*, April 5, 2005.

53. Allison North Jones, "College Bill Backer a Man of Conviction," *Tampa Tribune*, April 17, 2005.

54. James Vanlandingham, "Capitol Bill Aims to Control 'Leftist' Profs," *Independent Florida Alligator*, March 23, 2005; Diane Hirth, "Council Approves 'Academic Freedom,'" *Tallahassee Democrat*, April 20, 2005.

55. Scott Jaschik, "Academic Freedom Wars," *Inside Higher Ed*, March 25, 2005; James Vanlandingham, "Academic Bill Suddenly Reverses," *Independent Florida Alligator*, April 19, 2005.

56. Janine Young Sikes, "University Heads Hear Baxley's 'Burden,'" *Gainesville Sun*, April 22, 2005; Dennis Karp, "Lawmaker Takes His Complaints to the Top," *St. Petersburg Times*, April 22, 2005.

57. Alton Biggs, Whitney Crispen Hagins, Chris Kapicka, Linda Lundgren, Peter Rillero, and Kathleen G. Tallman, *Biology: The Dynamics of Life* (Columbus, Ohio: Glencoe/McGraw-Hill, 2004), 388.

58. Chris Kahn, "Choice of Biology Text Avoids National Debate," *Sun-Sentinel*, November 23, 2005.

59. Chris Kahn, "Biology Text May Stir Up Debate," *Sun-Sentinel*, November 27, 2005; Nikki Waller, "Textbook Bringing Creation Debate to S. Fla.," *Miami Herald*, December 3, 2005.

60. Hannah Sampson, "'Design' Passage Might Be Cut," *Miami Herald*, December 9, 2005; Ronnie Blair, "Teachers Opt for Texts without Intelligent Design," *Tampa Tribune*, January 29, 2006.

61. James Dean, "Textbook Debate Still Evolving," *Florida Today*, March 12, 2006.

62. School Board of Brevard County, Florida, Minutes of Regular Meeting, March 14, 2006, 5.

63. Ibid., 6.

64. Steven Ray Haberlin, "Intelligent Design Text in School Libraries," *Ocala Star-Banner*, November 3, 2005.

65. Jennifer Fennell, "Statement from Education Commissioner John L. Winn Regarding Florida's Sunshine State Standards for Science," press release, Florida Department of Education, October 11, 2005, http://www.fldoe.org/news/2005/2005_10_11.asp.

66. Katherine Lewis, "Creation vs. Evolution Debate Takes Center Stage," *Naples Daily News*, October 17, 2005; Donna Winchester, "Intelligent Design Makes for Big Bang"; *St. Petersburg Times*, November 5, 2005; Blair, "Intelligent Design Debate Brews."

67. Winchester, "Intelligent Design"; Ronnie Blair, "Intelligent Design Debate Brews," *Tampa Tribune*, October 12, 2005.

68. Donna Winchester, "Poll: Evolution's Not Enough," *St. Petersburg Times*, December 30, 2005.

69. Matthew I. Pinzur, "Evolution Showdown Delayed," *Miami Herald*, December 1, 2005; Ron Matus, "Questioning Florida's Science Standards," *St. Petersburg Times*, December 30, 2005.

70. Russell Schweiss, "Statement by Governor Jeb Bush Regarding Sunshine State Standards for Science," press release, Florida Department of Education, December 30, 2005, http://www.fldoe.org/news/2005/2005_12_30.asp; Jason Garcia, "Governor Sees Room to Debate Evolution," *Orlando Sentinel*, December 31, 2005.

71. Garcia, "Governor Sees Room to Debate Evolution."

72. "About Us," Physicians and Surgeons for Scientific Integrity, accessed October 4, 2012, http://www.pssiinternational.com/about.php.

73. Jacob Tillman, "Design Dilemma: Scientists Seek to Convince USF Audience of Intelligent Design's Validity," *University of South Florida Oracle*, October 2, 2006.

74. Mary Spicuzza, "Professor Crusades for Intelligent Design Concept," *St. Petersburg Times*, November 20, 2005.

Chapter 9. "I Want God to Be Part of This"

1. Eric J. Smith, "Approval of Amendment to Proposed Rule 6A-1.09401, Student Performance Standards," Florida Department of Education, February 19, 2008, http://www.fldoe.org/board/meetings/2008_02_19/Cover%201%20Science.pdf; Office of

Math and Science, "Why Do Florida's K–12 Students Need These New World-Class Science Standards?" Florida Department of Education, February 19, 2008, 5, http://www.fldoe.org/board/meetings/2008_02_19/FebBoardd.pdf; John Chambliss, "Science Standards: Plan to Require Evolution to Be Taught in Schools," *Lakeland Ledger*, October 24, 2007.

2. Office of Math and Science, "Florida's K–12 Students"; Joshua Rosenau, "Evolution Comes to Florida's Science Standards," *Reports of the National Center for Science Education* 28, no. 2 (2008): 4.

3. James A. Smith, "Fixing the Sub-standard Science Standards," *Florida Baptist Witness*, February 14, 2008; Marc Caputo, "Schools Await Board's Vote on Evolution," *Miami Herald*, February 18, 2008.

4. Casey Luskin, "Florida State Board of Education Receives Minority Report That Covers Evolution Objectively," *Evolution News and Views* (blog), February 5, 2008, http://www.evolutionnews.org/2008/02/florida_state_board_of_educati004805.html.

5. Office of Math and Science, "Florida's K–12 Students"; Mary Jane Tappen, "Florida's New World Class Education Mathematics Standards," Florida Department of Education, May 15, 2007, 17, http://www.fldoe.org/board/meetings/2007_05_15/Math%20Standards%20FINALb.pdf; Ron Matus, "Florida's Proposed Education Core Is More Specific and Now Includes Evolution," *St. Petersburg Times*, November 30, 2007.

6. Cary McMullen, "Opposing Sides Debate Change in Curriculum," *Lakeland Ledger*, November 16, 2007; John Chambliss, "School Official Opposes Evolution Standards Plan," *Lakeland Ledger*, November 13, 2007.

7. Bill Kaczor, "County Woman Leads Assault on Evolution Instruction," *St. Augustine Record*, December 9, 2007; Ron Matus, "Emphasis on Evolution Decried," *St. Petersburg Times*, December 6, 2007.

8. Ron Matus, "Moms against Darwin Dogma?" *St. Petersburg Times Gradebook* (blog), December 11, 2007, http://www.tampabay.com/blogs/gradebook/content/moms-against-darwin-dogma; Lisa Miller, "Difference Makers," *Florida Times-Union*, August 19, 2005; Mark Pettus, "A Trip to Dollywood Inspires a Literacy Project with the Eagle as Its Mascot," *Florida Times-Union*, October 6, 2006.

9. Michele Sager, "State Panel Skips Talk on Science Lessons," *Tampa Tribune*, December 12, 2007; James A. Smith, "Baptists Take Lead in Opposing Evolution-Only Standards," *Florida Baptist Witness*, December 20, 2007.

10. Smith, "Baptists Take Lead."

11. James S. Smith, "Sub-standard Science Standards," *Florida Baptist Witness*, December 6, 2007.

12. "Donna Callaway," Florida Board of Education, accessed October 5, 2012, http://www.fldoe.org/board/bios/Callaway.asp; Jacob DiPietre, "Governor Bush Appoints Donna G. Callaway to the State Board of Education," press release, Florida Department of Education, November 9, 2004, http://www.fldoe.org/news/2004/2004_11_09.asp; Smith, "Sub-standard Science Standards"; "Ignorance Has No Place in Curriculum," *St. Petersburg Times*, December 10, 2007.

13. Jeffrey S. Solochek, "State Rebukes Evolution Foe," *St. Petersburg Times*, December 8, 2007.

14. "State Board of Education: Where They Stand on Evolution," *St. Petersburg Times*, February 14, 2008; Matus, "Emphasis on Evolution Decried."

15. Chambliss, "School Official"; John Chambliss, "Majority Opposes Science Proposal," *Lakeland Ledger*, November 20, 2007.

16. Billy Townsend, "Polk Needled, Noodled in Evolution Flap," *Tampa Tribune*, December 22, 2007.

17. Fred Grimm, "Kids Paying for Holy War over Evolution," *Miami Herald*, January 10, 2008.

18. Ron Matus, "Several School Boards Say They Want to Teach Alternative Theories," *St. Petersburg Times*, January 24, 2008; "Those Not in Favor of Good Science Education, Raise Your Hand," *Florida Citizens for Science* (blog), February 15, 2008, http://www.flascience.org/wp/?p=352; Ira Mikell, "Evolution in Schools?" *Suwannee Democrat*, January 30, 2008; School Board of Nassau County, Florida, Agenda of Regular Meeting, January 24, 2008, 4; and Donna Valava, "County Wants to Teach More Than Evolution," *Northwest Florida Daily News*, February 14, 2008.

19. "Resolution," School Board of Baker County, December 17, 2007, http://www.flascience.org/BakerResolutionScience.pdf.

20. School Board of Putnam County, Florida, Minutes of Regular Meeting, February 19, 2008, 4; School Board of Highlands County, Florida, Minutes of Regular Meeting, February 5, 2008, 4; School Board of Monroe County, Florida, Minutes of Regular Meeting, February 12, 2008, 2; Pat Hatfield, "Volusia School Board Endorses Evolution," *DeLand-Deltona Beacon*, February 18, 2008.

21. Donna Winchester, "Four School Board Members Would Teach Intelligent Design Alongside Evolution," *St. Petersburg Times*, December 18, 2007; Matus, "Several School Boards."

22. "Science Storm Brewing?" *St. Petersburg Times Gradebook* (blog), November 29, 2007, http://www.tampabay.com/blogs/gradebook/content/science-storm-brewing; "Call to Action Project," *Florida Citizens for Science* (blog), December 6, 2007, http://www.flascience.org/wp/?p=337.

23. Elaine Silvestrini, "Candidate Calls Load 'Honest Mistake,'" *Tampa Tribune*, August 8, 2008; Adam C. Smith, "Conservative Isn't Retreating," *St. Petersburg Times*, May 19, 2007.

24. "Educators Say Evolution Still 'Theory,'" *World Net Daily*, February 19, 2008, http://www.wnd.com/2008/02/56830/.

25. Rebecca Steele, "Proposed Florida Science Standards," American Civil Liberties Union of Florida, December 14, 2007, http://www.aclufl.org/pdfs/BOE_LETTER.pdf.

26. Ron Matus, "Darwin Critics Arrive in Force," *St. Petersburg Times*, February 12, 2008; Janet Acerra, Fred J. Barch, Todd Bevis, Melody Boeringer, Susan D. Brennan, Kathryn Bylsma, and David Campbell to members of the Florida Board of Education, February 11, 2008, http://www.tampabay.com/blogs/gradebook/sites/tampabay.com.blogs.gradebook/files/images/typepad-legacy-files/48007.sboe8febfinal.doc; Hannah Sampson, "Fresh Airing of Evolution Draws Crowd," *Miami Herald*, January 9, 2008.

27. Harold Kroto, "Evolution Is a 'Theory' in Name Only," *St. Augustine Record*, February 16, 2008.

28. Ron Matus, "Gators, Bulls, 'Noles, 'Canes and Evolution," *St. Petersburg Times Gradebook* (blog), February 14, 2008, http://www.tampabay.com/blogs/gradebook/content/gators-bulls-noles-canes-and-evolution.

29. Ron Matus, "Lawyer Says School Proposal Equates Evolution, Religion," *St. Petersburg Times*, January 15, 2008; and Michael Stewart, "Lawyer: Hovind Detailed Actions: Evangelist Said He Beat the System," *Pensacola News Journal*, October 21, 2006.

30. David Gibbs to Linda Taylor, December 5, 2007, http://www.creationstudies.org/AR-BC320_20071205_110442.pdf.

31. Ron Matus, "Foster Links Darwin, Hitler," *St. Petersburg Times*, January 12, 2008; Adam C. Smith, "Can Creation Evolve into a Valid Issue?" *St. Petersburg Times*, September 16, 2009; Cristina Silva, "Foster Promises to Earn Respect," *St. Petersburg Times*, November 4, 2009.

32. Ron Word, "Public Debates How Florida Schools Should Teach Evolution," *Ocala Star-Banner*, January 4, 2008; Ron Matus, "Darwin Survives Another Debate," *St. Petersburg Times*, January 4, 2008.

33. Don Jordan, "Educators Support Evolution at Hearing," *Palm Beach Post*, January 9, 2008; Akilah Johnson, "Community Is Split on 'Evolution' in Schools," *Sun-Sentinel*, January 9, 2008.

34. Office of Math and Science, "Florida's K–12 Students."

35. James A. Smith, "Sub-standard Science Standards, Still," *Florida Baptist Witness*, February 7, 2008; Kaczor, "County Woman Leads Assault."

36. Matus, "Darwin Critics Arrive in Force."

37. Leslie Postal, "Science Hearing Gets 2 Viewpoints," *Orlando Sentinel*, February 12, 2008; Matus, "Darwin Critics Arrive in Force."

38. Tom Butler, "State Board of Education to Hear Public Testimony on Proposed Science Standards," press release, Florida Department of Education, February 14, 2008, http://www.fldoe.org/news/2008/2008_02_14.asp.

39. Leslie Postal, "4 Extra Words in Florida Science Proposal Rankle Educators," *Orlando Sentinel*, February 17, 2008.

40. Ron Matus, "Evolution of State's 'Theory' Rule Is Seen in Paper Trail," *St. Petersburg Times*, March 21, 2008.

41. Florida Department of Education, "February 19, 2008 Meeting Archive Part 1," video, accessed October 10, 2012, http://www.fldoe.org/board/meetings/2008_02_19/meetingArchive.asp.

42. Ibid.

43. Ibid.

44. Ibid.

45. Ibid.

46. Ibid.

47. Ibid.

48. Ibid.

49. James A. Smith, "Board Approves Science Standards with 'Theory' Compromise," *Florida Baptist Witness*, February 19, 2008.

50. Florida Department of Education, "February 19, 2008 Meeting Archive Part 1."

51. Ibid.

52. Ibid.

53. Ibid.

54. Ibid.

55. Florida Department of Education, "February 19, 2008 Meeting Archive Part 2," video, accessed October 10, 2012, http://www.fldoe.org/board/meetings/2008_02_19/meetingArchive.asp.

56. Ibid.

57. Ibid.

58. Ibid.

59. Ibid.

60. Ibid.

61. Ibid.

62. Ibid.

63. Ibid.

64. Ibid.

65. Ibid.

66. Ibid.

67. Ibid.

68. Ibid.

69. Leslie Postal, "Divided Florida Board Adopts Wording Compromise," *Orlando Sentinel*, February 20, 2008.

70. John Chambliss, "Evolution Controversy Settled, in Theory," *Lakeland Ledger*, February 20, 2008.

71. Bret Schulte, "Teaching Evolution in Florida," *U.S. News and World Report*, February 20, 2008.

72. Sylvia Lim, "Institute Criticizes Fla.'s 'Evolving' Standards," *Bradenton Herald*, February 22, 2008.

73. "Educators Say Evolution Still 'Theory,'" *WorldNetDaily*, February 19, 2008; Casey Luskin, "Florida State Board Tricked into Meaningless 'Compromise' to Retain Dogmatism and Call Evolution 'Scientific Theory,'" *Evolution News and Views* (blog), February 19, 2008, http://www.evolutionnews.org/2008/02/florida_state_board_of_educati_1004904.html.

74. Catherine Dolinski, "Critics Say Evolution Fight Not Over Yet," *Tampa Tribune*, February 20, 2008; Smith, "Board Approves Science Standards"; James A. Smith, "Rubio: Florida House Open to Legislative Fix on Evolution," *Florida Baptist Witness*, February 28, 2008.

75. Donna Callaway, "Students 'Biggest Losers' in Evolution Debate," *Florida Baptist Witness*, February 22, 2008.

76. Ibid.

Chapter 10. "Who Gets to Decide What Is Science?"

1. Bill Varian, "Mincing Few Words, Pulling No Punches," *St. Petersburg Times*, April 26, 2004.

2. Ibid.; Adam Smith, "Storms Makes Political Vow," *St. Petersburg Times*, April 28, 2006; "Taking Her Campaign to Talk Radio," *St. Petersburg Times*, January 18, 2006.

3. Alex Leary, "Lawmakers Want to Tax Strip Clubs to Aid Elderly," *St. Petersburg Times*, February 8, 2008; Dara Kam, "Christian License Plate Detoured in Senate," *Palm Beach Post*, April 26, 2008.

4. "SB 2692," Florida Senate, accessed October 16, 2012, http://archive.flsenate.gov/cgi-bin/view_page.pl?Tab=session&Submenu=1&FT=D&File=sb2692.html&Directory=session/2008/Senate/bills/billtext/html/.

5. Ibid.

6. "Support Academic Freedom," Academic Freedom Petition, September 7, 2007, accessed October 16, 2012, http://www.academicfreedompetition.com/freedom.php; Simon Maloy, "The Unscientific Model: 'Academic Freedom's' Creationist Pedigree," *Media Matters for America* (blog), April 17, 2012, http://mediamatters.org/blog/2012/04/17/the-unscientific-model-academic-freedoms-creati/184627; Jeffrey Solochek, "Storms Joins Darwin Debate," *St. Petersburg Times*, March 4, 2008; Marc Caputo, "Bill Could Allow Intelligent Design in Science Class," *Miami Herald*, March 12, 2008.

7. "Senator Alan Hays," Florida Senate, accessed October 16, 2012, http://www.flsenate.gov/Senators/s20; Jacob Ogles, "Open State House Seat Draws Ambitious Roster," *Daily Commercial*, August 12, 2004.

8. Terry Kemple, "Press Release: State Legislature to Take Up 'Academic Freedom' Act," *Community Issues Council*, March 1, 2008, http://www.tampabay.com/blogs/gradebook/sites/tampabay.com.blogs.gradebook/files/images/typepad-legacy-files/48223.22908_press_release_bill_filed.doc.

9. Robert Crowther, "Darwinist Activists at Florida Citizens for Science Think Academic Freedom Is 'Smelly Crap,'" *Evolution News and Views* (blog), March 5, 2008, http://www.evolutionnews.org/2008/03/darwinist_activists_at_florida004960.html.

10. James A. Smith, "Orlando Science Teacher Says Academic Freedom Bill 'Extremely Important,'" *Florida Baptist Witness*, April 1, 2008; Ron Matus, "Religion Has Nothing to Do with It?" *St. Petersburg Times Gradebook* (blog), March 5, 2008, http://www.tampabay.com/blogs/gradebook/2008/03/religion-has-no.html.

11. Ron Matus, "The Persecution Problem," *St. Petersburg Times Gradebook* (blog), March 6, 2008, http://www.tampabay.com/blogs/gradebook/2008/03/the-persecution.html; Catherine Dolinski, "Storms Tries to Put Evolution up for Vote," *Tampa Tribune*, March 4, 2008.

12. Matus, "The Persecution Problem"; "Florida Legislature Getting Expelled," *Florida Citizens for Science* (blog), March 7, 2008, http://www.flascience.org/wp/?p=497.

13. Marc Caputo, "Ben Stein Weighs In on Evolution Fight," *Miami Herald*, March 10, 2008; Bill Cotterell, "Legislators Invited to Private Screening," *Pensacola News Journal*, March 11, 2008; Whitney Ray, "Stein's Evolution," *WJHG*, March 12, 2008, http://www.wjhg.com/news/headlines/16620886.html.

14. Robert Crowther, "Press Conference March 12, 2008," audio mp3 files, accessed October 19, 2012, http://www.evolutionnews.org/2008/03/listen_to_ben_steins_comments004992.html.

15. Ibid.

16. Ibid.

17. Marc Caputo, "Bill Could Allow Intelligent Design in Science Class," *Miami Herald*, March 12, 2008.

18. Ibid.; "Florida Senate Bill 2692 Would Create Cover to Teach Religious Theories in Science Classrooms, Says ACLU," *ACLU*, March 12, 2008, http://www.aclu.org/religion-belief/design-should-not-be-taught-florida%E2%80%99s-public-school-science-classrooms.

19. Bill Cotterell, "Lawmakers Attend Tallahassee Screening of Movie by Ben Stein," *Tallahassee Democrat*, March 13, 2008.

20. Aaron Sharockman, "Bringing Evolution Controversy to the Capitol: Ben 'Bueller' Stein!?!," *St. Petersburg Times Gradebook* (blog), March 11, 2008, http://www.tampabay.com/blogs/gradebook/content/bringing-evolution-controversy-capitol-ben-bueller-stein; Brittany Benner, "Evolution Education Debate Continues," *WTSP*, March 18, 2008, http://www.wtsp.com/news/local/story.aspx?storyid=76331.

21. Michael Mayo, "'Academic Freedom' Bill a Way to Sneak Creationism into Schools," *Sun-Sentinel*, March 20, 2008; Mary Ann Lindley, "Some Simply Thrive on Power and Payback," *Tallahassee Democrat*, March 16, 2008.

22. Shannon Colavecchio-Van Sickler, "New Legislation to Keep Debate on Evolution Alive," *St. Petersburg Times*, March 13, 2008; Cotterell, "Lawmakers Attend Tallahassee Screening."

23. Anthony Man, "Consideration of Evolution Bill Is Questioned," *Sun-Sentinel*, March 26, 2008; John Chambliss, "Science Standards Options Proposed," *Lakeland Ledger*, March 5, 2008.

24. Professional Staff of the Education Pre-K–12 Committee, "Bill Analysis and Fiscal Impact Statement," Florida Senate, March 26, 2008, 2, http://archive.flsenate.gov/data/session/2008/Senate/bills/analysis/pdf/2008s2692.ed.pdf.

25. "Bill No. SB 2692 Committee Amendment," Florida Senate, March 26, 2008, http://archive.flsenate.gov/data/session/2008/Senate/bills/amendments_Com/pdf/sb2692AM398832.pdf; Professional Staff of the Education Pre-K–12 Committee, "Bill Analysis"; Professional Staff of the Judiciary Committee, "Bill Analysis and Fiscal Impact Statement," Florida Senate, April 7, 2008, 5, http://archive.flsenate.gov/data/session/2008/Senate/bills/analysis/pdf/2008s2692.ju.pdf.

26. Schools and Learning Council, "House of Representatives Staff Analysis," Florida House of Representatives, April 9, 2008, 4, http://www.myfloridahouse.gov/Sections/Documents/loaddoc.aspx?FileName=h1483.SLC.doc&DocumentType=Analysis&BillNumber=1483&Session=2008.

27. Laura Green, "'Evolution Act' Passed by Panel," *Palm Beach Post*, March 27, 2008.

28. Nicola M. White, "Evolution Dissent Advances," *Tampa Tribune*, March 27, 2008; Green, "'Evolution Act' Passed by Panel."

29. White, "Evolution Dissent Advances."

30. David Brackin, "We Need to Hear from Public School Science Teachers," *inJesus*, April 2, 2008, http://www.injesus.com/message-archives/teaching-education/CEAI-friends/ceai-action-alert-for-science-teachers; Christopher Collettee, "Evolution

Critics Get Behind Legislative Bill," *WTSP*, March 25, 2008, http://www.wtsp.com/news/local/story.aspx?storyid=76820.

31. Florida Senate, *Judiciary Committee Meeting April 8, 2008*, CD.

32. Ibid.

33. Ibid.

34. Ibid.

35. Ibid.

36. Ibid.

37. Ibid.

38. "Committee Substitute 1," Florida House of Representatives, April 11, 2008, http://www.myfloridahouse.gov/Sections/Documents/loaddoc.aspx?FileName=_h1483c1.xml&DocumentType=Bill&BillNumber=1483&Session=2008.

39. Florida House of Representatives, "House Schools and Learning Committee April 11, 2008," video, accessed October 16, 2012, http://www.flascience.org/video/041108_houselearncomm.wmv.

40. Ibid.

41. Ibid.

42. Ibid.

43. Ibid.

44. Ibid.

45. Ibid.

46. "Media Alert," *Florida Citizens for Science* (blog), April 10, 2008, http://www.flascience.org/wp/?p=535.

47. Anna Scott, "Evolution Fray Attracts Top Scientist," *Sarasota Herald-Tribune*, April 15, 2008.

48. James A. Smith, "Storms Faces Attacks over Evolution Bill," *Florida Baptist Witness*, April 14, 2008.

49. "Senator Amendment," Florida Senate, April 17, 2008, http://archive.flsenate.gov/data/session/2008/Senate/bills/amendments/pdf/sb2692c1376830.pdf.

50. Florida Senate, "Senate 2nd Reading April 17, 2008," video, accessed October 16, 2012, http://www.flascience.org/video/041708_senate2read.wmv.

51. Ibid.

52. Ibid.

53. Ibid.

54. Ibid.

55. Ibid.

56. Ibid.

57. Ibid.

58. Ibid.

59. Linda Kleindienst, "Lawmakers: Free Speech Protection Should Extend to Sex Ed Classes, Too," *Sun-Sentinel*, April 17, 2008.

60. Florida Senate, "Senate 3rd Reading and Vote April 23, 2008," video, accessed October 16, 2012, http://www.flascience.org/video/042308_senatefinal.wmv.

61. Ibid.

62. Ibid.

63. Ibid.

64. Ibid.

65. Ibid.

66. Ibid.

67. Florida House of Representatives, "House 2nd Reading April 25, 2008," video, accessed October 16, 2012, http://www.flascience.org/video/042508_house2read. wmv.

68. Ibid.

69. Ibid.

70. Ibid.

71. Ibid.

72. Ibid.

73. Ibid.

74. Ibid.

75. Ibid.

76. Ibid.

77. Ibid.

78. Florida House of Representatives, "House 3rd Reading and Vote April 28, 2008," video, accessed October 16, 2012, http://www.flascience.org/video/042808_housefinal.wmv.

79. Ibid.

80. Ibid.

81. Ibid.

82. Ibid.

83. Ibid.

84. Ibid.

85. Ibid.

86. Ibid.

87. Ibid.

88. Ibid.

89. Ibid.

90. Ibid.

91. Ibid.

92. Ibid.

93. Ibid.

94. Ibid.

95. Ibid.

96. Ibid.

97. Catherine Dolinski, "Vote Thins Evolution Bill's Odds," *Tampa Tribune*, April 29, 2008.

98. Adam Hasner, "Media Issue Brief: CS/SB 2692—Evolutionary Theory," *The House Majority Office*, n.d.

99. James A. Smith, "Senate, House Adopt Different Evolution Academic Freedom Bills," *Florida Baptist Witness*, May 1, 2008.

100. Steve Patterson, "Schools Evolution Proposal Could Die," *Florida Times-Union*, May 1, 2008.

101. "Dead," *Florida Citizens for Science* (blog), May 2, 2008, http://www.flascience. org/wp/?p=574; James A. Smith, "When No Action Is Good—and Bad—in Legislating," *Florida Baptist Witness*, May 6, 2008.

Chapter 11. "Standing Up for the Little Guy"

1. "Darwin Day & FCS Annual Meeting," *Florida Citizens for Science* (blog), January 30, 2008, http://www.flascience.org/wp/?p=414.

2. Daniel Ruth, "Evolution? Darwin Day? Big Problem," *Tampa Tribune*, May 8, 2008.

3. Lorena Madrigal, e-mail to author, September 6, 2011.

4. Nathan Crabbe, "Creationism and Evolution Debated," *Gainesville Sun*, October 28, 2008.

5. Matt Soergel, "Wise to Introduce Bill on Intelligent Design," *Florida Times-Union*, February 8, 2009.

6. Ibid.

7. Ibid.

8. Ron Matus, "'Why Would You Pass a Law That Would Invite a Very Expensive Lawsuit?'" *Gradebook* (blog), February 10, 2009, http://www.tampabay.com/blogs/ gradebook/2009/02/why-would-you-p.html; Ron Matus, "Evolution vs. Intelligent Design: The Tallahassee Battle Returns," *Gradebook* (blog), February 9, 2009, http:// www.tampabay.com/blogs/gradebook/2009/02/evolution-v-int.html.

9. Seán Kinane, "Scientists Oppose Intelligent Design Bill," WMNF Community Radio, February 23, 2009, http://www.wmnf.org/news_stories/6823.

10. "SB 2396," Florida Senate, February 27, 2009, http://archive.flsenate.gov/data/ session/2009/Senate/bills/billtext/pdf/s2396.pdf.

11. "'Critical Analysis' Bill Undermines Florida Science Education," *Florida Citizens for Science* (blog), February 27, 2009, http://www.flascience.org/wp/?p=926; Richard L. Turner, "A Position Statement from the Florida Academy of Sciences," *Florida Academy of Sciences*, March 20, 2009, http://www.flascience.org/fas_statement.pdf.

12. William March, "Anti-evolution Bill Still a Fruitless Exercise," *Tampa Tribune*, March 28, 2009.

13. Klint Lowry, "What's Old Is New Again," *Tampa Tribune*, April 18, 2009. Woodward continues to organize a variety of events based on the "Darwin or Design" theme featuring many nationally known intelligent design proponents. He has also written the books *Doubts about Darwin: A History of Intelligent Design* and *Darwin Strikes Back: Defending the Science of Intelligent Design*.

14. Ronnie Blair, "Teacher Workshop to Focus on Controversial Science Topics," *Tampa Tribune*, July 12, 2009.

15. Samantha R. Fowler and Gerry G. Meisels, "Florida Teachers' Attitudes about Teaching Evolution," *American Biology Teacher* 72, no. 2 (2010): 96–99.

16. Ron Matus, "Textbook Promotes Creationism, Florida Science Group Says," *Gradebook* (blog), September 22, 2010, http://www.tampabay.com/blogs/gradebook/ content/textbook-promotes-creationism-florida-science-group-says.

17. Leslie Postal, "Marine-Science Book Will Lose Passages That Prompted Criticism," *Sentinel School Zone* (blog), September 30, 2010, http://blogs.orlandosentinel.com/news_education_edblog/2010/09/marine-science-book-will-lose-passages-that-prompted-criticism.html; Leslie Postal, "Publisher Agrees to Cut 'Pro-Creationism' Material from High School Science Textbook, State Officials Say," *Orlando Sentinel*, September 23, 2010.

18. "Creationism Pops Up in Textbook Adoption Process," *Florida Citizens for Science* (blog), September 22, 2010, http://www.flascience.org/wp/?p=1227.

19. Leslie Postal, "Textbook Company Official: 'We Teach Evolution, and We Teach It to the Sunshine State Standards,'" *Sentinel School Zone* (blog), October 4, 2010, http://blogs.orlandosentinel.com/news_education_edblog/2010/10/textbook-company-official-we-teach-evolution-and-we-teach-it-to-the-sunshine-state-standards.html.

20. "SB 1854," Florida Senate, March 5, 2011, http://www.flsenate.gov/Session/Bill/2011/1854/BillText/Filed/PDF.

21. Elaine Silvestrini, "Evolution Challenge Has Some Alarmed," *Tampa Tribune*, March 13, 2011.

22. Ron Matus, "Florida Anti-evolution Bill Is 'Waste of Our Lawmakers' Precious Time,'" *Gradebook* (blog), March 14, 2011, http://www.tampabay.com/blogs/gradebook/content/florida-anti-evolution-bill-waste-our-lawmakers-precious-time; Abel Harding, "Senator: Teach Evolution Alternative," *St. Augustine Record*, March 16, 2011.

23. Joni B. Hannigan, "Answers in Genesis Raises Awareness Says Hibernia Youth Pastor," *Florida Baptist Witness*, September 15, 2011.

24. Michele Sager, "Olson Only School Board Incumbent to Win Race," *Tampa Tribune*, August 25, 2010; Marlene Sokol, "Race Has a Lot of Angles," *Tampa Bay Times*, October 21, 2012.

25. Peter Guinta, "House District 17 Candidates Present Agendas," *St. Augustine Record*, July 12, 2012.

26. Leslie Postal, "Florida Gets a D in Science," *Orlando Sentinel*, January 31, 2012; Lawrence S. Lerner, Ursula Goodenough, John Lynch, Martha Schwartz, Richard Schwartz, and Paul R. Gross, *The State of State Science Standards 2012* (Washington, D.C.: Thomas B. Fordham Institute), 46, http://www.edexcellencemedia.net/publications/2012/2012-State-of-State-Science-Standards/2012-State-of-State-Science-Standards-FINAL.pdf.

27. Erik W. Robelen, "Nationwide Review of Science Standards Offers Low Marks," *Education Week*, February 8, 2012.

28. Leslie Postal, "What Should FL Do about Its Science Standards? (The Ones Rated Poorly in National Study)," *Sentinel School Zone* (blog), April 4, 2012, http://blogs.orlandosentinel.com/news_education_edblog/2012/04/what-should-fl-do-about-its-science-standards-the-ones-rated-poorly-in-national-study.html.

29. School Board of Columbia County, Florida, Minutes of Regular Meeting, January 10, 2012.

30. Laura Hampson, "Merriken vs. Board of Ed," *Lake City Reporter*, March 11, 2012.

31. "Florida Tax Payers," advertisement, *Lake City Reporter*, April 15, 2012.

32. "To Candidates for Florida's Columbia County School Superintendent," advertisement, *Lake City Reporter*, June 24, 2012.

33. School Board of Columbia County, Florida, Minutes of Regular Meeting, December 11, 2012, 9; School Board of Columbia County, Florida, Minutes of Regular Meeting, January 8, 2013, 6; School Board of Columbia County, Florida, Minutes of Regular Meeting, January 22, 2013, 8; School Board of Columbia County, Florida, Minutes of Regular Meeting, February 12, 2013, 8; and School Board of Columbia County, Florida, Minutes of Regular Meeting, February 26, 2013, 7.

Epilogue

1. Lauren Roth, "Group That Gave Out Bibles in Orange Schools Welcomes Competition," *Sentinel School Zone* (blog), January 18, 2013, http://blogs.orlandosentinel.com/news_education_edblog/2013/01/group-that-gave-out-bibles-in-orange-schools-welcomes-competition.html; "Our Objectives," World Changers of Florida, accessed March 24, 2013, http://www.worldchangersfl.com/?page_id=12.

2. Zack Kopplin, "Creationism Spreading in Schools, Thanks to Vouchers," *MSNBC*, January 16, 2013, http://tv.msnbc.com/2013/01/16/creationism-spreading-in-schools-thanks-to-vouchers; Zack Kopplin, "Creationist Voucher Schools in Florida," *Say No to Creationist Vouchers* (blog), November 30, 2012, http://creationistvouchers.com/2012/11/30/creationist-voucher-schools-in-florida.

Index

BRANDON HAUGHT'S former jobs include Marine Corps combat correspondent, newspaper columnist, and graphic designer. He currently works as a sheriff's office spokesperson in Florida. He is also a founding board member and volunteer communications director for Florida Citizens for Science.

The University Press of Florida is the scholarly publishing agency for the State University System of Florida, comprising Florida A&M University, Florida Atlantic University, Florida Gulf Coast University, Florida International University, Florida State University, New College of Florida, University of Central Florida, University of Florida, University of North Florida, University of South Florida, and University of West Florida.